LOCALITIES

TITLES OF RELATED INTEREST

America's suburban centers
R. Cervero

The carrier wave
P. Hall & P. Preston

Cities and telecommunications
M. Moss

Cost-benefit analysis in urban and regional planning
J. A. Schofield

Distribution-free tests
H. R. Neave & P. L. Worthington

Gentrification of the city
N. Smith & P. Williams (eds)

Housing and urban renewal
A. D. Thomas

Intelligent planning
R. Wyatt

Landscape meanings and values
D. Lowenthal & E. C. Penning-Rowsell (eds)

London 2001
P. Hall

London's Green Belt
R. J. C. Munton

The politics of the urban crisis
A. Sills *et al.*

Property before people
A. Power

Remaking planning
T. Brindley *et al.*

Town and country planning in Britain
J. B. Cullingworth

Urban and regional planning
P. Hall

Urban problems in Western Europe
P. Cheshire & D. Hay

Western sunrise
P. Hall *et al.*

LOCALITIES

THE CHANGING FACE OF
URBAN BRITAIN

Edited by

Philip Cooke

*University of Wales
Cardiff*

London
UNWIN HYMAN
Boston   Sydney   Wellington

Published by the Academic Division of
Unwin Hyman Ltd
15/17 Broadwick Street, London W1V 1FP, UK

Unwin Hyman Inc.,
8 Winchester Place, Winchester, Mass. 01890, USA

Allen & Unwin (Australia) Ltd,
8 Napier Street, North Sydney, NSW 2060, Australia

Allen & Unwin (New Zealand) Ltd in association with the
Port Nicholson Press Ltd,
Compusales Building, 75 Ghuznee Street, Wellington 1, New Zealand

First published in 1989

British Library Cataloguing in Publication Data

Localities: the changing face of urban Britain.
1. Great Britain. Urban regions. Social change
I. Cooke, P. N. (Philip Nicholas), *1946–*
307.7'6'0941
ISBN 0–04–445300–0

Library of Congress Cataloging-in-Publication Data

Localities: the changing face of urban Britain/edited by Philip Cooke.
    p. cm.
Report from a research program organized by the
Economic and Social Research Council.
Bibliography: p.
Includes index.
ISBN 0–04–445300–0
1. Cities and towns – Great Britain – Growth – Case studies.
2. Regional planning – Great Britain – Case studies. 3. Great Britain –
Industries – Location – Case studies. 3. Urban renewal – Great Britain –
Case studies. I. Cooke, Philip (Philip N.)
HT384.G7L63 1989
307.7'6'0941 – dc19.   88–37071
                                                                CIP

Typeset in 10 on 11 point Bembo by Computape (Pickering) Ltd
and printed in Great Britain by Billing and Son Ltd, London and Worcester

Acknowledgements

This is the first book arising from the Economic and Social Research Council's Changing Urban and Regional System (CURS) research programme. Many researchers contributed to the programme and I wish to thank all of them for helping this book to reach fruition. For their invaluable secretarial and research assistance on the National Study, some of the results of which appear in the first chapter, I wish to thank Jean Rees, Nina Parry and Tod Rutherford. In technical terms our efforts were greatly assisted by the warmth and efficiency of Nigel Walford and his co-workers at the ESRC Data Archive, University of Essex, various staff at the Office of Population Consensus and Surveys at London and Titchfield, the Department of Employment Statistics Division at Watford and the NOMIS team at the University of Durham. Alec O'Neil, formerly of Sheffield University, was commissioned to provide the electoral time-series data and did so most effectively. Special thanks for written comments on one or other of the chapters which I wrote go to Martin Boddy, Harry Cowen, John Lovering, Richard Meegan and John Urry; from the CURS Steering Committee Doreen Massey, Derek Lyddon and Brian Robson; and, as a departmental colleague, Peter Williams. I am grateful to Howard Newby for diverse forms of help in the early stages of the programme, and in easing the path to publication.

Preface

In 1984 the Economic and Social Research Council established the Changing Urban and Regional System (CURS) research programme. I was appointed coordinator of the programme in October of that year, with responsibility, in addition, for undertaking a study of the spatial impact of social, economic and political change on the UK. Seven teams were selected in March 1985 to undertake intensive studies of localities chosen for their variety of experience in the context of the broader national and international changes identified in the study. The research programme came to an end in December 1987. The overall objectives of the programme were to explore the impact of economic restructuring at national and local levels, and to assess the role of central and local government policies in enabling or constraining localities, through their various social and political organizations, to deal with processes of restructuring. The work grew out of the research fellowship, funded by the Social Science Research Council, awarded earlier to Professor Doreen Massey, to investigate the changing industrial geography of the UK with a view to generating explanations of the processes observed. The 'restructuring thesis', which informed the CURS programme to a significant extent, arose in large measure from the fellowship findings.

Although stimulated by interest in and concern about the effects of economic restructuring, it was the aim of CURS researchers not to be limited simply to exploring economic processes. Indeed, an important dimension of the research involved seeking to establish the conceptual status of the idea of 'locality' by taking account of a wide range of social scientific theory and research. This mode of thinking not only informed the approach to seeking to understand the complexity of social interactions at the local level and between it and larger scales, but was also reflected in the teams themselves, whose members were drawn from a multitude of disciplinary backgrounds.

The policy dimension of local interaction with the regional, national and supranational levels was an important motivator of the research, too, given both changes in the relative strengths of central and local government powers and the often difficult problems of adjustment to new opportunities or misfortunes localities were confronting. The findings of the research are thus addressed as much to the policy community and interested laypersons as to the academic community, and the style of this book is intended to reflect this.

The research programme was organized throughout a steering committee consisting originally of Professor Brian Robson as chairperson, Professor Doreen Massey, Professor Howard Newby, Professor Noel Boaden, Dr Duncan Gallie and myself. The administrative assistance to the steering committee was provided by Dr Angela Williams, followed by Dr John Malin and Mr Kevin Hamilton. The coordinatorship and the national study were funded by ESRC grant number D04250001. The coordination of the programme and the national study were carried out from the Department of Town Planning of the University of Wales, Cardiff. The seven locality study teams were drawn primarily from the University of Durham, the University of Lancaster, Centre for Environmental Studies (CES) Ltd, the University of Aston, the Gloucester College of Advanced Technology, the University of Bristol, the University of Essex and the University of Kent at Canterbury.

Cardiff, 1988   PHILIP COOKE
Coordinator, ESRC Changing
Urban and Regional System
Research Programme

Contents

Contributors

Paul Bagguley Lancaster Regionalism Group, Department of Sociology, University of Lancaster, Lancaster LA1 4YW

Keith Basset Department of Geography, University of Bristol, Bristol BS8 1SS

Huw Beynon Department of Sociology, University of Manchester, Manchester M13 9PL

Martin Boddy School for Advanced Urban Studies, University of Bristol, Bristol BS8 4EA

Nick Buck Urban and Regional Studies Unit, University of Kent at Canterbury, Canterbury CT2 7NZ

Philip Cooke Department of Town Planning, University of Wales, Cardiff CF1 3YN

Harry Cowen School of Environmental Studies, Gloucester College of Advanced Technology, Gloucester GL50 2RR

Ian Gordon Urban and Regional Studies Unit, University of Kent at Canterbury, Canterbury CT2 7NZ

Michael Harloe Department of Sociology, University of Essex, Colchester CO4 3SQ

Steve Harrison School of Environmental Studies, Gloucester College of Advanced Technology, Gloucester GL50 2RR

Laurie Howes School of Environmental Studies, Gloucester College of Advanced Technology, Gloucester GL50 2RR

Ray Hudson Department of Geography, University of Durham, Durham DH1 3HP

Brian Jerrard School of Environmental Studies, Gloucester College of Advanced Technology, Gloucester GL50 2RR

Jim Lewis Department of Geography, University of Durham, Durham DH1 3HP

Ian Livingstone School of Environmental Studies, Gloucester College of Advanced Technology, Gloucester GL50 2RR

John Lovering School for Advanced Urban Studies, University of Bristol, Bristol BS8 3EA

Jane Mark-Lawson Lancaster Regionalism Group, Department of Sociology, University of Lancaster, Lancaster LA1 4YW

CONTRIBUTORS

Andy McNab   School of Environmental Studies, Gloucester College of Advanced Technology, Gloucester GL50 2RR

Richard Meegan   Centre for Environmental Studies Ltd, 5 Tavistock Place, London WC1

Chris Pickvance   Urban and Regional Studies Unit, University of Kent at Canterbury, Canterbury CT2 7NZ

David Sadler   Department of Geography, St David's University College, Lampeter

Dan Shapiro   Lancaster Regionalism Group, Department of Sociology, University of Lancaster, Lancaster LA1 4YW

Dennis Smith   Management School, University of Aston, Birmingham B4 7ET

Peter Taylor-Gooby   Urban and Regional Studies Unit, University of Kent at Canterbury, Canterbury CT2 7NZ

Alan Townsend   Department of Geography, University of Durham, Durham DH1 3HP

John Urry   Lancaster Regionalism Group, Department of Sociology, University of Lancaster, Lancaster LA1 4YW

Sylvia Walby   Lancaster Regionalism Group, Department of Sociology, University of Lancaster, Lancaster LA1 4YW

Alan Warde   Lancaster Regionalism Group, Department of Sociology, University of Lancaster, Lancaster LA1 4YW

Locality, Economic Restructuring and World Development

PHILIP COOKE

Introduction

This book sets out to answer a difficult question about life in the late twentieth century. While people's lives continue to be mainly circumscribed by the localities in which they live and work, can they exert an influence on the fate of those places given that so much their destiny is increasingly controlled by global political and economic forces? The most important version of this question in the 1980s has concerned prospects of survival in the face of a second Cold War between the USA and USSR. This threatened to heat up into the most dangerous conflict possible on more than one occasion in the past decade as the two superpowers struggled to gain advantage in the balance of military and political power. Such was the popular alarm at these posturings that new kinds of social movement, such as the Women's Peace Protest in Britain, the Nuclear Freeze campaign in the USA, and European Nuclear Disarmament (END), came into existence to join rejuvenated movements such as the Campaign for Nuclear Disarmament (CND).

These movements are not, in the sense of this book, local in their scope, though they would not exist in practical terms without often very strong local or regional organizational bases. Indeed, one of the issues which is addressed in subsequent chapters is why social protest campaigns such as the Peace Movement or the Women's Movement are stronger in some localities than others and whether such spatial variation is explicable in general terms. Are there, in other words, locally rooted social processes capable of projecting the interests of locality members well beyond the local political arena, and what might those processes be? Moreover, why should some localities be good at mobilizing local interests and others more acquiescent towards the outside world? Why is the evenly spread presence of democratic rights unevenly acted upon? Why do some issues anger people in certain localities more than people in others, and why do

1

Figure 1.1 The seven study localities.

certain places tend to be associated with specific types of popular mobilization?

To try to answer these and related questions about local variety the book focuses on the more pervasive experience of economic restructuring, social adjustment and political change by examining how seven very different localities have been coping with the vagaries of Britain's long-term industrial decline. Specifically, it is concerned with the ways in which economic restructuring has interacted with the study localities over the period beginning in 1970. The localities in question are Middlesbrough, Lancaster, outer Liverpool, south-west Birmingham, Cheltenham, Swindon and Thanet. The impact of economic change may be direct and can be brutal in its effects on individuals and households. It can ruin the hopes of families in this or the next generation, or offer opportunities for people to redirect their lives by breaking out of a particular groove, changing roles, switching jobs, or moving to a more prosperous town.

Already the word 'locality' has been mentioned many times as a simple, descriptive term for the place where people live out their daily working and domestic lives and around which they may, on occasion, act more politically than simply voting in local, national or, for the EEC, supranational elections. The chances are that it is the lcoal area within which residents receive hospital treatment, children go to school, many leisure activities are pursued and views are exchanged about matters of common interest, from the state of the roads to the state of the nation. In the next section there will be a more extended discussion of the meaning of the term and an attempt to clarify the way it is used in the rest of the book.

The central aim of the book is to examine the extent to which localities can act as a viable base for social mobilization and exert influence upon outside forces which help shape their destiny. To do this it is necessary to examine the nature of those forces which, recently as in earlier periods, have disrupted the economic basis of all localities – creating opportunities for some as it closes livelihoods down for others, in a process to which the economist Joseph Schumpeter once referred as 'creative destruction' (Schumpeter, 1943, p. 81). As Schumpeter saw, a key force is capitalism, which

> ... is by nature a form or method of economic change and not only never is but never can be stationary ... that incessantly revolutionizes the economic structure *from within*, incessantly destroying the old one, incessantly creating a new one. This process of Creative Destruction is the essential fact about capitalism (Schumpeter, 1943, pp. 82–3, emphasis in original).

3

Capitalism has changed since Schumpeter wrote about it. Generally, it has penetrated deeper into the national economies of the world as a method of production as well as simply a form of exchange. Specifically, in the case of the UK, it has relinquished its former strongholds of production in favour of new locations in developing economies while retaining a still dominant financial base in London. In the process many localities have lost their economic rationale while many others find they have a new one, perhaps as the 'global outpost' (Austrin and Beynon, 1979) of a multinational corporation or the electronic abacus of a City of London finance house. Hence, the third section of this chapter examines the ways in which key national economies have been coming to terms with the changing relationships of the international economy. In doing this, attention is drawn to the main themes making up this changed and extended global setting.

The spatial implications of these global shifts of the 1970s and 1980s for the UK are traced in the fourth section. By showing detailed patterns of change at locality level it is possible to gain some indication of the spatially uneven distribution of advantage and disadvantage as it has developed. The most important employment location tendencies in the 1970s were metropolitan deindustrialization and the urbanization of the countryside (Figures 1.2 and 1.3). Substantial shifts of population and employment from cities were in part an effect of the failure by older, city-based UK firms to compete with those from overseas. But movements of this kind may also be an expression of changed preference in residual location, in this case for what may be perceived as a 'rural idyll'. In the past the process of suburbanization that produced the urban sprawl of the interwar years arose as jobs followed residential development, especially when employers realized that the suburbs contained sizeable pools of relatively cheap, female labour (Hobsbawm, 1968; Mills, 1973; Scott, 1982). As Scott (1982) points out, though, while a multiplicity of 'factors' are at work in the decentralization process only a few of these can be seen in operation using information about employment change. To be interested in the changing nature of localities as distinct from local economies involves tracing changes in their *social* composition and, for example, the extent to which private as against public consumption of services, density of white or blue collar occupation, or level of professional qualification may have varied from place to place over time. In turn, this may suggest potentially important developments in local *political* allegiance with, perhaps, as Edgell and Duke (1986) suggest, evidence of newly developing relationships between voting practices and occupational structure across the public-private employment divide.

4

Figure 1.2 Percentage change 1971–1981 in manufacturing employment (Great Britain) and the secondary sector (Northern Ireland). Source: OPCS.

Lastly, the chapters on the seven selected localities will be introduced. Three main aspects of these localities in the context of the overall research programme require stressing. First of all, their degree of 'representativeness' or otherwise warrants consideration. Clearly, to take seven localities as being in any sense statistically representative of the 334 (UK) travel-to-work areas comprising the localities defined in this study, would be inappropriate. But it would be wrong to

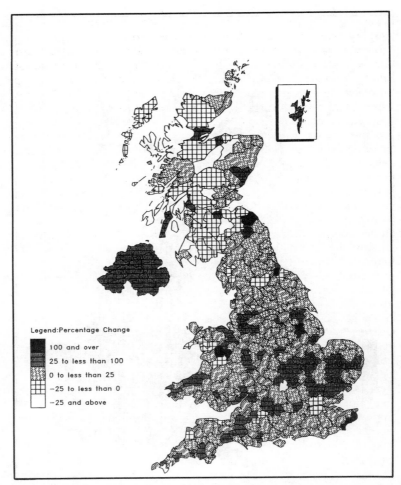

Figure 1.3 Percentage change 1971–1981 in distribution and other services employment (Great Britain) and the tertiary sector (Northern Ireland). Source: OPCS.

suggest that the kinds of locality studied were chosen by purely random processes. The research task was to examine the effects of economic restructuring on localities and, in turn, the impact of localities on restructuring processes. This meant that selection was influenced by theorizations of restructuring processes.

The main strategies identified with restructuring (Massey and Meegan, 1982; Massey, 1984) are as follows. Firms confronted with a severe competitive challenge can rationalize both capacity and employment, or intensify labour processes to achieve greater output from the same workforce, or undertake an investment programme, for example updating technology. These strategies can be adopted singly or in combination. To re-examine and develop these propositions in the new period of restructuring, the 1970s and 1980s, it was essential to select localities with industries that had been forced to respond to heightened competition. Thus, local economies experiencing change in services such as tourism or in manufacturing activities such as vehicle engineering were strong candidates for selection. Conversely, localities which were booming in the teeth of recession would obviously be of great interest. These two criteria governed the selection process, as did the notion that it would be important to examine anomalous cases, notably the prospering 'northern' locality and the beleaguered 'southern' one. The second dimension of the selection grid was that localities should have dissimilar combinations of local socio-economic and political characteristics (for further detail see Cooke, 1986a).

Second, the localities constitute an integrated set of diverse places of which comparable questions have been asked. More is known about the same range of issues for this group of places than for any other in the UK. Despite their diversity, some common as well as distinctive dimensions of change have been observed, especially with respect to responses to the twin economic pressures of heightened competition in the private sector and the widespread application of performance indicators in the public sector. Thus, guided by the general restructuring framework noted above, it had been possible to explain certain industrial and spatial shifts that occurred in the 1960s and early 1970s. There had been a considerable growth of new, decentralized office and manufacturing activities in localities that had taken their employment character from very different activities. Parts of Merseyside, Tyneside, Clydeside and South Wales, hitherto known for their heavy industry and port-related functions, became locations for central government organizations such as the Post Office Giro Bank, Department of Social Security, and Business Statistics Office as government sought sites with cheaper labour and land rents. With these moves new technologies were often installed to modernize data and other information-processing tasks.

Such spatial shifts were even more pronounced in manufacturing. Many older firms closed their doors in the inner city, sometimes as a result of merger or takeover activities encouraged by government. New vehicle, electrical and mechanical engineering investment took place in the more depressed conurbations or in medium-sized towns and semi-rural settings. Overall, manufacturing employment declined as firms rationalized, but the rate was greatest in the inner areas of the conurbations. This process of decentralization brought new workers into the labour market such as women part-timers. Established male full-time workers were accordingly in a more precarious position. Even in centres of skilled engineering work, such as the Midlands, automation made inroads into what had been secure male employment. Concentration of ownership meant that research, executive and administrative activities were likely to be relocated to headquarter locations increasingly found in and around London. Hence, the restructuring of the 1960s and 1970s produced a new emphasis in the economic geography of the UK: a 'spatial division of labour' (Massey, 1984) between an executive centre in London, a weakened craft belt in the semi-peripheral conurbations of the Midlands and the north, and a periphery characterized by growing employment in routine, semi-skilled factory and office work, partially compensating for the loss of jobs in heavy industry.

However, in the 1980s new developmental pressures were beginning to place that spatial division of labour under strain. The emergence in the late 1970s of ferocious overseas competition led to something of a reversal of the decentralization process as engineering and office branches were closed. Non-price competition grew with the result that the search for cheap labour was to some extent replaced by the quest for minimum defects and high-quality design. Part of this requirement has been met by a growth in subcontracting (OECD, 1986). Instead of accepting a standardized product or service aimed at a mass market, the increasingly discerning consumer could command more diversified output from what were becoming more segmented markets (Cooke and Morgan, 1985, 1986). To achieve economies of scale with diversification firms had to embark on large-scale investments in advanced manufacturing technology, 'paperless offices' and, in retailing, electronic point-of-sale equipment capable of rapidly relaying market signals to producers (Rajan, 1987). The development of new products and services required greater integration of design, management and marketing tasks, whereas these had often been physically separated previously. Labour force requirements were increasingly for functionally flexible workers capable of undertaking tasks requiring multiple skills and numerically flexible workforces available to work on short-term contracts as and when required (Atkinson, 1984). Economic organization has thus become more flexible under this latest phase of restructuring, representing a move away from what have been called 'Fordist' to 'post-Fordist production methods (Piore and Sabel, 1984). 'Fordism' was associated with large-scale, mass production methods as pioneered by Henry Ford in Detroit and extended even to the factory building of mass housing in the 1960s. 'Fordism' is a shorthand for the method of social, political and economic regulation which linked, by making interdependent, mass production and mass consumption.

In brief, the spatial trend in the 1980s is towards centralization in or near the urban settings where flexible suppliers and workforces are more common and where older managerial styles, often equated with conflict-ridden relations with unionized workers, are absent. Such localities tend to be Southern and suburban or semi-rural, except for the highest level service activities which remain metropolitan. Accordingly specific localities may have remarkably different profiles in their direction and rate of employment change in the major industry groupings. This clearly relates to location, local social history and capacity to take advantage of the changing field of opportunities. Many elements of the flexibility just discussed are observable in the seven localities studied.

However, it is important to comment on the extent and nature of possible generalizations that can be made about changes in the wider space economy on the basis of the locality studies. Some rather wide-ranging deductions can be drawn with reasonable confidence where the official UK-wide secondary data is being interpreted. But it would be wrong to expect a severely limited sample to allow representative generalizations to be drawn from the locality evidence. Rather, coherent and well-informed hypotheses or propositions (such as the one about 'flexibility') can be advanced where this research seems to agree broadly with the findings of related research undertaken elsewhere. Some of these may take the form of preliminary theoretical constructs awaiting further testing by research. In one respect only do we think our generalizations can be regarded as rather strong, and this is that each locality study has sought to generalize *within* rather than *across* cases. That is, the research programme from which these findings arise has involved seven research teams investigating the ways in which localities function from the same broad theoretical perspective and asking many similar questions derived from that perspective. The extent to which localities function similarly in the face of a common, though multifaceted, experience of economic restructuring clearly cannot be prejudged. But whatever the answer it will, inasmuch as it rests on the study of seven very different places, be one of the stronger general conclusions of the whole research programme.

Locale, locality and community

There is a gap in the social science literature when it comes to a concept dealing with the sphere of social activity that is focused upon place, that is not only reactive or inward-looking with regard to place, and that is not limited in its scope by a primary stress on stability and continuity. Community is inadequate because it fails to satisfy the second and third conditions. Also, it fails to satisfy the first condition in that it is not strictly denotative of place. It is by no means unusual to speak of community aspatially as in the idea of communities of interest or 'community without propinquity' (Webber, 1964). Community is, therefore, too broad in its spatial reach and too narrow in its social connotations, especially in respect of its external content. 'Locality' is a strong candidate for filling the gap.

Before considering the adequacy of *locality* to the task in hand it is worth briefly considering the value of the apparently cognate concept of *locale* recently introduced by Giddens (1984). As Giddens puts it,

> Locales refer to the use of space to provide the *settings* of interaction, the settings of interaction in turn being essential to specifying its contextuality ... Locales may range from a room in a house, a street corner, the shop floor of a factory, towns and cities, to the territorially demarcated areas occupied by nation-states (Giddens, 1984, p. 118).

There are three reasons for rejecting the notion of locale as a candidate to fill the gap. The first of these is that it is even looser than community in its spatial scope. For, although there are such appellations as the 'community of nations' and, even more formally, the European Community, even common usage seldom extends its application to the scale of the room. The second reason for rejecting locale is that it reproduces the passive connotations of community in the way it refers to setting and context for action rather than as a constituting element in action. Finally, unlike even the inadequate notion of community embodied in the work of earlier theorists of community, locale lacks any specific social meaning. It remains primarily a synonym for space – a particularly old-fashioned one, since it stands for a space which can be *occupied* by socio-political entities such as the nation state.

A better conceptualization, offered by Savage et al. (1987), sees locality as the product of the interactions of supralocal structures. These may, it is proposed, give rise to local specificity, but this never goes as far as to warrant the designation 'local culture' though it does merit the appellation 'locality'. However this approach is too restrictive in its structural determination. It shares with community and locale a blind spot with respect to the active, potentially effective power embodied within both the concept of locality and the practices of its members. If locality is reduced to the interactive outcome of common structural determinations it becomes impossible to explain local variations between otherwise similarly constituted places. Yet we know that historical and contemporary social practices of an innovative character emerge in specific localities, sometimes in more than one more or less simultaneously.

Historical examples would include the ways in which industrial districts such as Sheffield, Stoke-on-Trent, Limoges, Lyons and Bradford emerged either singularly or sometimes in parallel as localities displaying particular productive specialisms. Outside the industrial sphere, the development of localities with particular aesthetic specialisms such as Salzburg for opera, Cannes for film appreciation and Nashville for country and western music cannot readily be explained without considerable reference to local initiatives of a collective kind. Returning to industry and the highly contemporary period new specialisms are being forged in small towns such as Treviso in Italy where clothing production is dominated by, but by no means limited to, the activities of the Benetton company. Meanwhile, Boston in Massachusetts has become an important software and high technology centre and Cambridge (UK) is blossoming as a research and development complex in industrial science and engineering.

Such examples probably have to be understood in terms that include but move beyond the purely structural. Each is a clear illustration of local mobilization by a few or many individuals and groups taking full advantage of what may be called *pro-active capacity*. What this means is rather straightforward. In sovereign nation states there are two main levels at which collective identification is actively expressed. The first is the national level: individuals are officially accorded a nationality with which as subjects they identify to a greater or lesser degree. The second is the local level where individuals work and live their everyday lives for the most part. At both these levels individuals of the appropriate age have democratic rights and freedoms of expression. Furthermore, the existence of such rights implies, in practice, the existence of the two levels. One of the key attributes of sovereignty is the right to *name* localities, a fact made visible whenever the control of territory is transferred as, for example, when colonizing powers annexe it or when emergent nations win independence. Membership of a locality by birth, or by residence for a qualifying period, admits the individual to the citizenship bestowed on 'nationals' of the soveriegn power.

Citizenship is thus a precondition for participation in the range of affairs that may take place in the territory of the nation state. Those lacking citizenship and residing in an alien nation state lack many of the instruments of pro-activity such as welfare or electoral rights, though they may possess certain economic rights. Such groups constitute and fall back upon what is more accurately called their community than their locality, though it is not unreasonable to consider such a community as existing in a particular locale.

Thus the discussion of the status and meaning of locality returns us to the point of departure. Citizenship as a means of social participation implies the existence of the modern nation state, an institution which became generalized only from the period when cultural modernity and industrial capitalism became dominant social paradigms. The modern nation state is one in which citizenship applies at both the local and the central levels. Both levels are key means for social mobilization and political intervention. For individuals who are *subjects* of the sovereign power of the nation state their citizenship rights are exchanged for their allegiance. Citizen subjects obtain civil, political and social rights (Marshall, 1977) in exchange for certain obligations such as those involving obedience to the law, including acceptance of military conscription in time of war. As Turner (1986) puts it: 'To be a citizen is to be a person with political rights involving liberty and protection in return for one's loyalty to the state' (Turner, 1986, pp. 106–7).

That the state can enforce this relationship is the result of modern methods of surveillance (Giddens, 1984) such as the various certificates, licences and numbers by means of which the modern state *locates* individuals in their localities.

In conclusion, locality is a concept attaching to a process characteristic of modernity, namely the extension, following political struggle, of civil, political and social rights of citizenship to individuals. Locality is the space within which the larger part of most citizens' daily working and consuming lives is lived. It is the base for a large measure of individual and social mobilization to activate, extend or defend those rights, not simply in the political sphere but more generally in the areas of cultural, economic and social life. Locality is thus a base from which subjects can exercise their capacity for pro-activity by making effective individual and collective interventions within and beyond that base. A significant measure of the context for exercising pro-activity is provided by the existence of structural factors which help define the social, political and economic composition of locality. But the variation between similarly endowed localities can only be fully understood in terms of the interaction between external and internal processes spurred, in societies dominated by capitalist social relations, by the imperatives of collective and individual competition and the quest for innovation.

The global agenda

Having stressed the importance of locality to the daily lives of individuals in modern society it is now necessary to consider some structural issues which set the agenda for local – and national – debate and manoeuvre. It has been necessary to stress citizenship as the source of pro-activity in the local and national contexts because it has often been neglected in debates of this kind. However, the agenda for the last quarter of the twentieth century also requires consideration of the phenomenon of 'global citizenship'. The idea is superficial in that there is no global state which can grant civil, political or social rights, although international attention and pressure are increasingly focused upon members of the community of nations falling below standards such as those set in, for example, the Helsinki agreement between the USSR and the USA. Moreover one supranational economic entity with pretensions towards political unity, the European Community, is displaying state-like characteristics in respect of, in particular, the freedom of movement of goods and people (in the case of the latter, Community-wide passports have been issued to replace national ones).

These are signs of a growing internationalization of economic and cultural, if not political affairs. Indeed internationalization could be said to be one of the key economic strategies adopted by industrial and financial capital to escape the crises of Fordism discussed earlier (see Castells and Henderson, 1987). The postwar era has been distinguished by an accelerating process of international investment. It is sometimes possible to assume, wrongly, that capital exports are headed down a one-way street marked Newly Industrializing Countries (NICs) and Less Developed Countries (LDCs), leaving the older industrial regions and cities of the developed world deindustrialized. Much has been made of the 'new international division of labour' thesis (Fröbel, Heinrichs and Kreye, 1980; for a critique see Hill, 1987) according to which foreign competition forces Western corporations to decentralize production to NICs and LDCs where labour is extremely cheap. This strategy allows the labour intensive parts of the production of commodities such as clothing, electronics, books and motor vehicle components to be cheapened sufficiently to allow companies to regain a competitive edge. This process is aided by technological developments in shipping, notably containerization, which have dramatically cheapened long distance transport. However, it is now becoming clear that such a strategy has only accounted for a small proportion of total international investment. By far the largest amount is reciprocally invested within the developed economies (Lipietz, 1982), and, in any case, there are signs that real wages are on the increase in NICs where the new international division of labour has been most marked, such as Hong Kong and South Korea.

Perhaps most important of all during the post-1975 period was not the obvious economic impact of the increased oil prices upon world inflation but the deteriorating political relations between West and East as the 1970s closed. The Soviet Union, then strongly influenced in its foreign policy by military interests, deployed medium-range nuclear missiles in Europe, invaded Afghanistan and was seen to be supportive of successful independence movements in Africa, South East Asia and Central America. The USA was, simultaneously, in a weakened position in world affairs following defeat in the Vietnam war and the loss of influence in Indochina more generally. Increases in tension in the Middle East did little to ease US foreign problems, a factor seriously aggravated by a total loss of influence in Iran. The Nicaraguan revolution only underlined a perception among political conservatives that the 1970s had marked a postwar nadir in American political influence overseas. The election of Ronald Reagan as President in 1980 signified the ascendancy of the New Right to political power and the commitment to military, monetary and moral strength at home and overseas. The articulation of a strident anti-Soviet rhetoric backed up with a major increase in armaments expenditure and the deployment of medium-range nuclear missiles in Europe bore witness to US bellicosity. Localities tied to the American arms build-up, especially those connected with the $6 billion US 'Star Wars' programme, have been strengthened economically by defence contracts. But many also became the focus for protest from a resurgent international Peace Movement.

A second major international tendency of the 1980s has been the more or less coordinated effort by the larger Western economies to control inflation. The rise of monetarist economic theory, which blamed the appearance of double-digit inflation on what were perceived as excessive increases in public welfare expenditure 'crowding out' more efficient 'productive' private investment, influenced conservative governments to cut welfare programmes and raise interest rates prohibitively. Even though such policies have, by the late 1980s, been largely discredited for their failure actually to reduce overall levels of public spending due to the increase in unemployment partly caused by their deflationary effect upon industry, inflation has been brought under control. However, localities in which public spending on housing, health and other local services was proportionately high due to inherited economic, hence tax-raising, difficulties, have been particularly badly hit by the combination of a further deindustrialization and loss of revenues. Moreover, those localities dependent on

14

state-owned and state-subsidized industries such as shipbuilding, steel manufacture and coal mining have been exposed to massive job loss by national and international (e.g. EEC) policies aimed at returning them to profitability and private ownership.

Thirdly, there has been a fluctuating political, industrial and financial struggle between the developed and the less developed countries, particularly centred upon the crucial issues of the price of oil. The 1970s saw the quadrupling of oil prices by the OPEC countries, many of which were underdeveloped by Western standards. While this (and the economic effects of Vietnam) had the greater effect on inflation, in turn causing serious problems for the weaker developed economies, in the long run it forced Western countries to become more energy-efficient. The rate of increase in consumption of oil therefore declined at approximately the same time that monetarist policies were raising interest rates to record levels. Accordingly, those oil-rich developing countries that had borrowed heavily from Western banks using oil as collateral in order to invest in industrial development, found themselves caught in the pincers of declining oil revenues and increasing debt charges. Debt restructuring and virtual write-offs have become common practices on the part of creditor companies, weakening some banks and forcing them into international mergers in some cases. Regions and localities in developed countries producing oil or controlling its production have experienced a switchback of fortune as boom conditions led to investment in heavy engineering, shipyard work and oil-related financial services, and slumps produced industrial downturns and job lay-offs. Those cities in oil-debtor countries that had been experiencing rapid development are now stagnating and the austerity policies forced on debtor governments to finance payments have caused tremendous social hardship followed by increased rural–urban and international migration. It has been estimated that legal immigration into the US alone may be as high as 880,000 per year with a further half a million entering illegally each year in the 1980s (Cohen, 1987).

A fourth element in the global framework in the recent past, one which is more purely economic than some, is the relative decline in the rate of productivity growth in manufacturing industry as compared to that of the period leading up to 1973. The date is no coincidence: it has recently been argued that the major reason for lower productivity growth in the advanced economies after 1973 was that firms were forced to spend capital on higher-cost energy that might otherwise have purchased more efficient plant and machinery. As a result the relative average rates of productivity increase were as shown in Table 1.1. The main means of raising productivity have been redundancy and, in the US at least, real wage reductions by comparison with the

Table 1.1 *Average rates of manufacturing productivity growth*

	Manufacturing 1950–1973	Productivity (%) 1973–1983
Japan	10.0	6.8
West Germany	6.5	3.7
France	5.8	4.6
UK	3.3	2.4
US	2.8	1.8

Source: Marshall (1986)

1976 peak. In the UK, where unemployment has been between 1 and 6 percentage points higher than that in the US between 1979 and 1986 (Green, 1986) real wages have been rising during the 1980s. Thus labour productivity in the UK has been bought by means of unemployment funded, it can be argued, by North Sea oil revenues. The extent to which the latter have been contributing to the UK economy in the recent past is shown in Table 1.2.

Capital productivity has been raised in manufacturing by the restructuring of business organization, including investment in new production methods and technologies. As a result, manufacturing productivity growth in the UK, taking into account both labour and capital productivity gains, has been double that of the total economy from 1981/2 to 1984/5. As in the US the absorption of labour into low-order and miscellaneous service sector activities acts as a drag on overall productivity. Other strategies in pursuit of productivity gains have included the export of capital to cheap labour zones and its redeployment into higher yielding sectors, including financial markets.

Fifth, the rise of a number of Newly Industrializing Countries has posed a challenge to the competitive dominance of the advanced economies in certain key economic sectors. Consumer electronics producers in the West have been unable to compete with products from South East Asia and Japan, and in such product markets American and European companies have either retired from the contest or, as in the case of video cassette recorders, hardly developed a presence. Clothing is a major and growing market in which NICs are competitive. The motor industry is similarly though not as significantly under threat from such NICs as South Korea and Taiwan, Japan being obviously the major competitive presence for the foreseeable future. Some NICs in Latin America and the Middle East as well as South East Asia have become major producers in heavy industries such as steel and petrochemicals. While price competition

16

Table 1.2 *Profit rate trends in the UK economy 1980–1985*

Year	Profit rate[1] of UK including oil	Economy (%) excluding oil
1980	6	3
1981	6	3
1982	7	4
1983	9	5
1984	12	7
1985	13	8

[1] Defined as profits net of consumption as a proportion of total equity.
Source: Green (1986)

in a world steel market, for example, that has stagnated at 700 million tonnes per year since 1974 had been causing large reductions in market share for traditional steel producing countries such as the UK, industrial restructuring has made such producers more competitive in terms of quality. Thus far competition from NICs in services is less pronounced, though Singapore at $9.5 billion in 1982 was the ninth largest exporter of services in the world (Noyelle, 1989).

A sixth global tendency of the 1980s has been the growth in migration from the less developed countries to the more developed, including intra-LDC movement. The three major migration streams have been: first, that of LDC residents to the global cities of the US and Europe; second, that of LDC residents to NICs, and particularly to their export processing zones; and third, that of migrants to the oil producing countries of OPEC. The latter are overwhelmingly from the poorer countries but some of these migrants include workers from regions of developed countries which have been deindustrialized but whose skills still have market value in industrializing countries (Sassen-Koob, 1984, 1987). With respect to migration to global cities such workers are filling occupations the growth in number of which is stimulated directly by demand for miscellaneous services and some lower order office employment. This mirrors the growth of higher-order managerial and professional employment, mostly in services and particularly producer services. The latter have been boosted by the growth in world trade, banking, insurance, finance and related stock exchange activities. Incomes can be excessively high in these services, especially during a deregulatory period: accordingly, the ability to pay for the consumption of a multitude of personal, domestic and other services is enhanced. Additionally, though, migrants from LDCs may be driven away from their normal country of residence by economic difficulties consequent upon the debt burdens and associated reductions in employment opportunities in countries seeking to develop industrially. Many such workers end up in the growing 'sweatshop' work in the clothing and small-scale engineering industries of the inner city. Such informal, small-firm and homeworking activity may, in some cases, be linked to the formal economy as larger firms seek ways of implementing cost-cutting, flexible production methods, in part by the subcontracting of work.

Seventh, and last, internationalization involves the widening and deepening of markets. This is a two-way process in which Western markets are increasingly extended, but also penetrated, protectionist rhetoric notwithstanding, by low-cost LDC producers. For example, it has recently been estimated that the world recession of 1980–82 caused an 8.9 per cent average GNP reduction in Latin America with some countries suffering as much as a 13–16 per cent loss. The general response to this by LDCs was to seek to limit trade imbalances by means of export drives. These resulted in an increase of Third World manufactured exports to the OECD from $60.5 billion in 1979 to $108.8 billion in 1984 (Portes and Sassen-Koob, 1987). However, trade flows into new, large and developing markets continue to increase as firms in the developed economies seek out new ways of ovecoming problems of market saturation in their home countries. During the same period OECD countries' trade with LDCs has also risen as countries with very large populations, such as China, India and Brazil especially, are perceived as important consumption as well as production locations. To this should be added the growth in traded services between the dominant financial centres such as London, New York and Tokyo and growing NIC and LDC cities such as Hong Kong, Singapore, Bombay and São Paulo. Much of the growth in global finance activity reflects the growth in the international division of labour within manufacturing. But it also signifies growth in international market niches for specialist financial services such as commodities, insurance and currency exchange. Moreover, inter-nationalization has increasingly become a policy of governments committed to the deregulation of stock market and related services in order to expose them to the astringent effects of more open market competition. Employment, business and income growth in leading world finance centres such as London are inducing substantial dis-tortions into labour and housing markets well beyond metropolitan boundaries.

These major dimensions of global political and economic power comprise a framework of enablements and constraints for the unfold-ing of capitalist development processes within specific nation state territories such as the UK. Moreover, as an economy that has traditionally been remarkably open to foreign trade, and dependent for its survival upon export markets, the UK may be said to be a particularly clear example of a country experiencing the extremes of global restructuring, both negative and positive, as the world economy shifts its axis in directions which undermine historically powerful regions and present opportunities for development to hitherto insignificant ones.

The changing urban and regional system in the UK

By the early 1960s, the urban and regional system in the UK was in transition from a 'pre-Fordist' spatial structure to a more fully developed Fordist one. The main developmental thrust was one in which the regional systems inherited from the Victorian era were to a large extent being replaced by a national system in which London had attained overwhelming importance as the city in which were centred the conception, control and executive functions of a growing proportion of industrial, commercial and governmental activities of relevance to UK cities and regions. This may seem odd given London's traditional role as a capital city. However the concentration and centralization of capital that occurred in the 1960s and 1970s (Westaway, 1974) meant that London and some subsidiary centres in the south-east of England were developing centralized control more widely and deeply than ever before.

There are numerous indicators, some circumstantial, of the effects of the centralization of control upon the urban system of London and the south-east. First, between 1974 and 1982 floor space occupied by commercial offices increased by 1.6 million square metres in central London alone, accounting for just under one-third of the total in the south-east where it grew by 29 per cent during the same period (Wood, 1987). It has been shown, too, that in the City, St Marylebone and Croydon districts of London there was employment growth between 1978 and 1981, and that in central London as a whole higher-order private sector services employment continued to grow while higher-order public administration, transport and communications employment bore the brunt of the declining component of services employment. Moreover, jobs in manufacturing industries grew by 5.3 per cent (10,100) in central high-employment areas over the same period, a shift only explicable in terms of the growth in central London of higher-level controlling functions in manufacturing industry (Simmie, 1985). Lastly, it should be noted from Figure 1.8 that the growth in the employer, manager, professional and intermediate socio-economic groups ranged up to 20 per cent in the London travel-to-work area between 1971 and 1981. This was not as high as the rate of growth in the rest of the south-east which exceeded 40 per cent in some localities, but it is supporting evidence for the general point that London's control functions were being enhanced. It is, finally, worth bearing in mind that all of the growth discussed occurred in a context where, as Marquand (1983) shows, Greater London had by 1971 easily the highest managerial and professional location quotients in the country.

However, during the period when the grip exerted by London-headquartered companies upon the UK space economy was tightening, it is clear also that the growth of London as a base for foreign capital was accelerating. It has become clear as a result of the work of Thrift (1987) and Thrift, Leyshon and Daniels (1987) that the growth in London of foreign producer services companies has been explosive since 1970. In that year just under 12,000 people were employed in 163 foreign banks, for example, but by 1984 employment had more than tripled to just under 40,000 working in 566 foreign banks and securities houses. In this respect London remained well in advance of New York as the major foreign banking centre in the world measured by the number of companies present. By 1986, the year when City institutions were deregulated in the 'Big Bang', there were 400 foreign *banks* alone in London employing approximately 54,000 – a sign of the startling build-up of employment in firms that had established footholds in readiness for opening up the hitherto domestically dominated financial markets (Thrift, 1987b).

This readjustment of London's role – once essentially a British-run centre controlling national and international spheres of commercial and financial activity, now an international base of the first importance in the regulation of global markets for banking and other financial exchange activities – echoes the wider processes of internationalization in economic affairs more generally. In short, London's affluent now reap a tremendous benefit from its burgeoning global role, in terms of further income and wealth generation. In turn these benefits spill over into the surrounding south-eastern towns and cities, none of which is remotely credible as a competitor, but many of which can play a subaltern role to London's command functions. Accordingly, growing numbers of south-eastern, East Anglian and south-western localities have themselves shown substantial proportional increases in employment in producer services (banking, finance, insurnce, etc.) during the 1970s and early 1980s (Figure 1.4). As Table 1.3 shows, the five free-standing localities in the present study all showed near or above average employment growth in producer services from 1981 to 1984, but the southern ones showed the greater numerical and often percentage growth. Also, the sharp growth of male part-time employment in producer services nationally

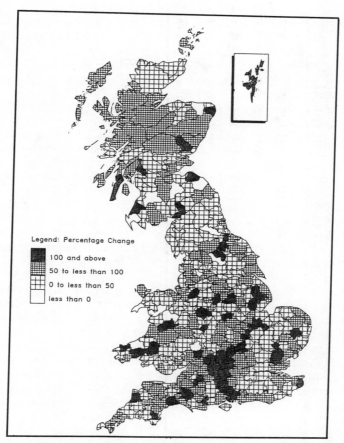

Figure 1.4 Percentage change 1971–1981 by locality – all in employment producer services (financial and professional). Source: D.E. Statistics (NOMIS) 1984 travel-to-work areas.

and in Thanet should be noted as a possible indicator of an emergent *numerical flexibility* in this growing set of industries.

These very limited comparisons between specific parts of the United Kingdom's urban and regional system begin to provide some insight into the nature of the changes being experienced differentially by the diverse types and locations of town and city found within it. The dominant southern labour market of London has been experiencing substantial employment growth in employer, manager and professional occupations, especially in producer services. Such has

21

Table 1.3 *Employment change in producer services[1] 1981–84*

	Thanet 1981 1984 (%)	Swindon 1981 1984 (%)	Cheltenham 1981 1984 (%)	Lancaster 1981 1984 (%)	Middlesbrough 1981 1984 (%)	GB (%)
Female, part-time	219 255 (16.4)	511 701 (37.2)	541 708 (30.6)	278 395 (42.1)	1092 1312 (20.1)	(17.7)
Female, full-time	670 800 (19.4)	2665 3331 (25.0)	2468 2731 (10.7)	669 799 (19.3)	1949 2392 (22.7)	(15.0)
Male, full-time	511 1021 (99.7)	2603 3846 (47.7)	2701 3128 (15.8)	642 830 (29.3)	2589 2927 (13.1)	(13.5)
Male, part-time	47 112 (72.3)	83 65 (−21.7)	87 24 (−72.4)	71 59 (−17.0)	92 99 (7.6)	(42.4)
Total	1447 2118 (46.4)	5862 7943 (35.5)	5797 6591 (13.7)	1660 2083 (25.5)	5722 6730 (17.6)	(14.5)

[1] Defined as 1980 SIC Activity Descriptions 8140–8395
Source: Census of Employment NOMIS Statistics

been the effect of this labour market expansion that it has contributed to a reversal of London's recent history of population loss. According to Champion (1987) inner London has, since 1983–84, been gaining population largely through a massive decline in outward migration (including a surplus of international inward migrants) and, to a lesser extent, a growing excess of births over deaths. The populations of outer London also grew, contributing to an average annual increase of 0.1 per cent in London as a whole from 1983 to 1985.

This growth does not appear to have been at the expense of small to medium-sized southern towns and cities, many of which have also experienced sharp employment growth, especially in producer services. The emphasis on this latter employment sub-category is warranted by the fact that it is unquestionably a major factor in local economic dynamism in the UK urban and regional system, and one furthermore that distinguishes good from poor economic performance within it. An interesting but presently unanswerable question concerns the nature and extent of linkages between producer services growth in central London and that in the Home Counties (Figure 1.5). A phenomenon recently noted in American city regions of large metropolitan scale has been the intra-corporate decentralization of 'back-office' work, such as data processing, from central city headquarters locations to suburban and exurban ones (Nelson, 1986). Evidence from this research programme points to the growth of employment in regional branch offices and the decentralization of headquarters or divisional offices *en bloc* as more likely explanations.

So far there is little evidence that the prosperity enjoyed by localities in the south is spreading more widely to other parts of the urban and regional system. That is not to say that the system has broken down: the opposite is the case, particularly in the sense that the control exerted over local economic and political life in the regions by corporations and government departments centred in London has grown. The UK urban and regional system is more and more integrated but it is an integration of opposites, a system based upon contradiction. Inspection of Table 1.4 shows the difference between southern and northern parts of the system for Great Britain. All of the localities with the highest long-term unemployment in 1985 lay outside the prospering regions of southern Britain and most of those with the lowest rates lay within an hour's travelling time of London. The most common characteristics shared by the depressed localities is some connection, however tenuous that may now be, with heavy industry such as dock work, coal mining, steelmaking or shipbuilding. Localities such as these northern and western centres of high long-term unemployment exist in a contradictory relationship to the prospering ones for many reasons, three of which are funda-

Figure 1.5 Percentage change 1981–1984 – all in employment producer services. Source: D.E. Statistics (NOMIS) 1984 travel-to-work areas.

mental. First, because control of local and regional labour markets was lost, the major employers in the locality became dependent on decisions taken elsewhere. In particular, their capacity to stimulate local industry through developing local subcontracting networks was diminished and an 'employee culture' resulted. This was reinforced by the existence of long-established and well-organized mechanisms for the defence of employee rights, wages and conditions in the form of trade unions, collective bargaining, seniority rules and so on. Although few of these localities could be described as classically Fordist, their indusrial histories shared, often having predated,

Table 1.4 *Highest and lowest long-term (>1 yr) unemployment*

Highest Rate (%)				Lowest Rate (%)			
1985		1987		1985		1987	
Middlesbrough	12.2	Cumnock	12.9	Crawley	1.4	Crawley	0.8
Cumnock	11.7	Liverpool	9.9	Shetland	1.4	Winchester	1.0
Hartlepool	11.4	S.Tyneside	9.8	Winchester	1.4	Basingstoke	1.1
S. Tyneside	11.3	Middlesbrough	9.7	Aberdeen	1.5	Bicester	1.1
Liverpool	10.8	Girvan	9.6	Aylesbury	1.7	Aylesbury	1.2
Sunderland	10.8	Hartlepool	9.3	Basingstoke	1.7	Cambridge	1.2
Irvine	10.7	Holyhead	9.2	Clitheroe	1.7	Guildford	1.2
Holyhead	10.4	Rotherham	9.2	Guildford	1.7	Newbury	1.2
Aberdare	10.2	Cardigan	9.1	Cambridge	1.8	Andover	1.3
Cardigan	10.2	Greenock	9.0	Andover	1.9	Bury St.E	1.3
		Irvine	9.0			Tunbridge. W	1.3

Source: Department of Employment (JUVOS) statistics)

Fordist characteristics such as labour deskilling, economies of scale, output standardization and task fragmentation.

Second, collective strength such as that displayed by widespread trade union consciousness and affiliation was reflected in a form of politics in which pressure was put upon the state apparatus through the Labour Party for mass provision of housing, health, education and welfare services. Such policies were seen to be superior means of universal social provision to those associated with the allocative mechanisms of the market. Accordingly, neither in the sphere of consumption – including that of food and clothing which were in part cooperatively distributed – nor in that of production (for many of the dominant industries in these localities had, by the 1970s, been nationalized) could the hegemony of private property ownership, entrepreneurship or an 'enterprise culture' be said to be more than vestigial.

Hence, thirdly, such localities would tend to be heavily dependent upon public expenditure and could be said, objectively, to be places in which the values of sociability, community, egalitarianism and social justice figured more prominently in political debate than questions of competition, monetary value, unit costs and performance indicators in the public sector. Issues of need would normally take precedence over those of affordability. However such social democratic impulses have fallen profoundly out of step with the dominant political mood of the 1980s. Tragically, localities in which these assumptions were most strongly rooted have experienced crippling reductions in employment from industries forced to compete on supposedly equal terms with overseas challengers. A similar fate has befallen housing

Table 1.5 *Population and employment change in large northern cities*

	Population			Employment		
	(1) 1961–1971	(2) 1971–1981	(3) 1981–1985[1]	(4) 1961–1971	(5) 1971–1981	(6) 1981–1984[2]
Birmingham	1.1	−5.2	−1.3	−2.5	−13.3	−3.7
Manchester	−1.4	−4.9	−2.5	−8.7	−10.6	−5.8
Liverpool	−3.8	−8.7	−4.9	−11.8	−14.2	−9.5
Glasgow	−4.1	−9.6	−5.2	−8.5	−16.5	−7.4
Newcastle	−5.7	−5.7	−0.7	−1.9	−8.5	−2.8
Sheffield	5.3	−6.1	−1.6	−2.0	−10.6	−9.6
Leeds	−2.9	−4.6	−1.0	−7.0	−7.0	−2.6

[1] Not strictly comparable with Columns 1 and 2 because OPCS & RG (Scotland) mid-year estimates
[2] Not strictly comparable with Columns 1 and 2, 4 and 5 because Census of Employment and travel-to-work areas
Sources: Columns 1 and 2, 4 and 5, Begg, Moore and Rhodes (1987)
Column (3), Champion (1987)
Column (6), Census of Employment (1984), NOMIS statistics

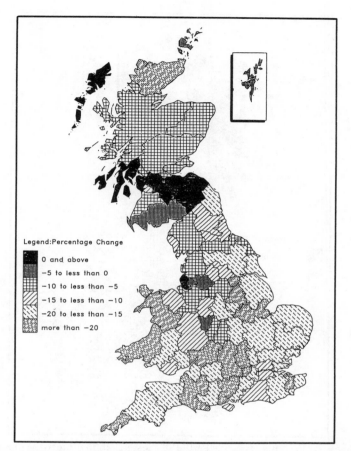

Figure 1.6 Percentage change 1970–1983: swing to Labour from Conservatives. Source: CURS General Election Time Series by 'Superconstituencies'.

services, with privatization and ratecapping preventing new building and repairs. Lastly, general welfare and educational services have suffered from the combination of cuts and penalties imposed by an increasingly parsimonious central state under fiscally austere neoconservative control.

Thus few northern localities fit the post-Fordist paradigm with its key elements of flexible production methods, privatized modes of ownership and service delivery, and increased competition in the various dimensions of social and cultural life. This is despite the considerable presence of large cities and conurbations throughout the

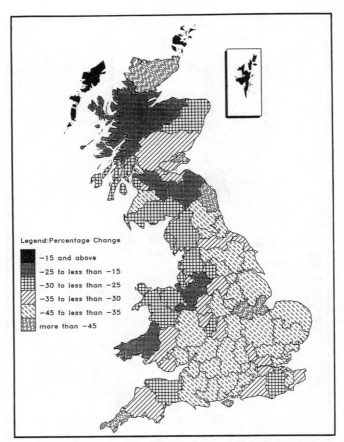

Figure 1.7 Percentage change 1970–1983: loss in Labour vote to the Liberals. Source: CURS General Election Time Series by 'Superconstituencies'.

north which once, but decreasingly, possessed key financial and control resources necessary for the economic and social restructuring occasioned by developments in overseas competition. As shown in Table 1.5, large northern cities have, up to 1984–5, been losing population and, most strikingly, employment in substantial proportions. However, it should also be noted that although the annual rate of population loss in most of these cities is less than it was in the 1970s the annual rate of employment decline was significantly higher.

The political expression of this contradiction within the UK urban and regional system shows up particularly clearly when the changes in share of the vote earned at general elections by the main British political parties are examined (Figures 1.6 and 1.7). If comparisons of this kind are to be made over a reasonable period of time it is necessary to aggregate parliamentary constituencies into larger constituency-based units, the boundaries of which change less than those of constituencies themselves. This is usually because certain administrative boundaries such as those of cities or counties form the template for groups of constituencies. These larger units or 'superconstituencies' also enable comparisons of electoral trends to be drawn with data assigned to groups of travel-to-work areas since they both have the ward as their basic spatial building block.

Between 1970 and 1983 the Labour Party's vote in general elections declined as a percentage of total votes cast in all superconstituencies. The Liberals were the greater beneficiary of changing allegiances, but the variation in the percentage shares achieved in the contest between Labour and Conservative provides the clearer picture of political polarization in Britain. It can be seen from Figure 1.6 that Labour lost significantly less of its share of the vote to Conservative in the north, and that in some parts of Scotland and on Merseyside Labour gained votes from the Tories. At a more local level the shift to Conservative was least in the large northern cities, especially Birmingham and Manchester, while in Liverpool there was a gain in votes. However, it is noticeable that the loss was less pronounced in the cities and region of north-west England than in those of the north-east and Yorkshire.

In the south, and for these purposes Wales can be included in this category, there was a very general change in the difference between the percentage shares of Labour and Conservative between 1970 and 1983 in favour of the latter. Within the broader picture it is worth noting the three kinds of areas where the shift away from Labour has been greatest. First is the block of constituencies based on small to medium-sized towns and cities stretching westwards from between Slough and Oxford to Cheltenham and southwards to Southampton, with an outlier in Somerset and Dorset. Here Labour's share of the vote declined by more than 20 per cent against that of the Conservatives over the 1970–1983 period. Second, there is a smaller area in the East Midlands west of the Wash where a similar Tory gain in share occurred. This area stretches northwards from south Lincolnshire and east Leicestershire to Doncaster in Yorkshire along what has become an extended tentacle of the London commuter belt. Third, Labour lost more than 20 per cent of its support by comparison with the Conservatives along the western seaboard and centre of Wales, and in Cornwall, a sign that Labour's tenuous grip on rural Britain had been all but relinquished. Elsewhere in the south the Tories skimmed off between 15 and 20 per cent of what had in 1970 been

Labour's share of the vote, except in the London to Brighton axis where Labour lost under 15 per cent of its vote to the Tories.

The pattern of decline in share of support for Labour in favour of the Alliance (until 1983 Liberals only) is rather similar in general terms to that in favour of the Tories. Scotland, the north-west of England and Wales registered the lower percentage shifts from Labour between 1970 and 1983. Southern, eastern and north-eastern England showed the largest percentage increases in Alliance share of the vote with Sussex, parts of Kent and parts of the south-west of England joining the cities with slightly more moderate Alliance gains.

It is evident, therefore, that the urban and regional system exists in dynamic tension as between the north and west of the country and London with its subaltern towns and cities, where there have been sizeable increases in the proportion of producer service jobs, employer, professional and managerial occupations, and Conservative and Alliance electoral gains from Labour in the 1970s and 1980s. The large cities of northern England and Scotland, which might be expected to play a leading role in the revival of regional economies, continue to lose population and employment. Many of the specialist industrial towns that developed fastest in Victorian times are now amongst the weakest labour markets in the country. Those towns that are in the north and prospering tend to be small and blessed with cultural or heritage resources (Stratford, Harrogate, Leamington) or a successful large employer (Aberdeen, Macclesfield, Stafford). The cities of the north and in particular the whole of north-west England and Scotland are relatively less affected by the shift away from supporting the Labour Party than the rest of Britain.

In the 1987 general election this contradictory, polarizing electoral tendency was reinforced except that the Alliance fell back everywhere with the other two major parties differentially the gainers. In north-west England Labour's percentage of the vote rose by more than 5 per cent while in Scotland and Wales it increased by over 7 per cent. In Yorkshire and the north-east the increase in Labour's vote was between these figures. Conservative support slipped by 2–3 per cent in the north. But in the south-east of England, and in Greater London particularly, the Tories increased their share of the vote by almost double that of Labour. In the East Midlands and south-west, Labour's increase in share was double that of the Conservatives. Labour strengthened in the northern cities, many of which now have no Tory representation in Parliament, but performed poorly in London (Massey, 1987). Hence the trend towards 'two nations' has by no means been reversed politically, any more than it has in social and economic terms (Morgan, 1986; Johnston and Pattie, 1987; Cooke,

1987). If anything the 1987 general election entrenched two apparently incompatible perspectives of the future in spatial and social terms. The one, a competitive, entrepreneurial and privatistic world view and social project for the future, the other, a more collectivist, socially conscious and fraternal perspective which remains locally powerful but on the defensive against the burgeoning state power of its ideological opponent.

Local accounts, local specificities

Within the 'two nations' stereotype – a prospering south, driven by London's international financial, cultural, tourist-attracting and governmental power, and a deindustrialized north, lacking many of these economic stimulants – there are exceptions. Moreover, different localities within the separate halves of the country demonstrate distinctive development profiles, showing that the combinations of locally specific features, whether economic, social, political or cultural, interact variably as particular developmental opportunities are opened or closed. But local distinctiveness is nevertheless confronted with powerful structural forces which can exert, at the most general level, markedly homogenizing effects. This is most obvious in two general characteristics of the national labour market as it has developed from 1971 to 1984. The first of these has been the precipitous relative and absolute decline in manufacturing occupations in the UK, and the second has been the steadily rising proportion, both relative and absolute, of employment to be found in the service sector.

In Table 1.6 it can be seen that the local development paths have shown less homogeneity in manufacturing employment change than in services. Those localities such as Middlesborough and Swindon that possessed very high manufacturing employment shares in large-scale processing industries such as chemicals, steel, and vehicle manufacture have seen those workforces cut drastically to close to the national share. Those localities where manufacturing employment was well below average suffered large proportional job reductions with the result that they remained as different from the mean in 1984 as they had been fifteen years earlier. And Cheltenham's developmental profile has been much out of step with its received image. It was, in 1984, returning to a ratio of manufacturing to services jobs more typical of itself in the early 1970s or of Britain in 1981. Surprisingly, Cheltenham had in 1984 a substantially greater percentage of its workforce engaged in manufacturing than Middlesbrough and only 6000 less manufacturing jobs despite the much greater size of the latter locality.

Table 1.6 *Manufacturing, services and total employment change, 1971–1981, 1981–1984*

(1968 SIC)	1971 (000s)	Share of Total (%)	1981 (000s)	Share of Total (%)	Change 1971–1981 (%)	1981 (000s) (1980 SIC)	Share of Total (%)	1984 (000s)	Share of Total (%)	Change 1981–1984 (%)
G.B.										
Manuf.	7,886	(36)	5,968	(28)	(−24)	6,253	(29)	5,337	(26)	(−15)
Services	9,805	(45)	11,545	(55)	(+18)	12,088	(57)	12,790	(61)	(+6)
Total*	21,638		21,092		(−3)	21,306		20,846		(−2)
SWB										
Manuf.	54	(60)	37	(43)	(−32)	37	(43)	33	(41)	(−11)
Services	30	(33)	42	(49)	(+40)	42	(49)	43	(53)	(+2)
Total*	90		85		(−6)	85		81		(−5)
Chelt'hm										
Manuf.	18	(32)	22	(30)	(+22)	21	(30)	20	(31)	(−5)
Services	31	(55)	43	(60)	(+39)	41	(59)	38	(59)	(−7)
Total*	56		74		(+32)	69		65		(−6)
Swindon										
Manuf.	32	(45)	21	(27)	(−34)	22	(27)	23	(27)	(+5)
Services	30	(42)	45	(59)	(+50)	48	(59)	54	(63)	(+13)
Total*	71		78		(+10)	82		86		(+5)
Lancstr										
Manuf.	11	(26)	10	(23)	(−9)	9	(21)	7	(16)	(−22)
Services	23	(53)	26	(59)	(+13)	27	(63)	27	(63)	(0)
Total*	43		44		(+2)	43		43		(0)

	No.	(%)	No.	(%)	Change	No.	(%)	No.	(%)	Change
O.L'pool										
Manuf.	35	(59)	21	(45)	(−40)	21	(45)	19	(45)	(−10)
Services	19	(32)	21	(45)	(+5)	21	(45)	19	(45)	(−10)
Total*	59		47		(−20)	47		42		(−11)
Thanet										
Manuf.	11	(27)	8	(24)	(−27)	8	(24)	6	(19)	(−25)
Services	22	(54)	19	(59)	(−14)	20	(61)	20	(65)	(0)
Total*	41		34		(−17)	33		31		(−6)
Middlesbro'										
Manuf.	55	(44)	35	(32)	(−36)	37	(34)	26	(27)	(−30)
Services	50	(40)	56	(52)	(+12)	58	(53)	62	(63)	(+7)
Total*	125		108		(−14)	110		98		(−11)

*Totals also include Primary, Utilities, Transport and Construction SICs

Source: Census of Employment NOMIS Statistics

Indeed, it is in terms of manufacturing employment growth that a distinction can be made between the two prospering southern localities of Swindon and Cheltenham and the rest. Swindon's manufacturing workforce grew after 1981, having slumped considerably in the 1970s. Cheltenham's grew markedly in the 1970s and only experienced a 6 per cent decline in the bad recession years of the early 1980s compared with a 15 per cent decline nationally. Cheltenham's manufacturing workforce was larger in 1984 than in 1971, a somewhat unusual feat for a UK locality. The other study localities lost on the whole greater percentages than those observed nationally during the 1971–81 and 1981–84 periods.

A further notable trend differentiating some study localities in the mid-1980s from their situation in the mid-1970s concerns the relative sizes of their manufacturing, services and total workforces. This is most apparent in the case of Thanet and Lancaster in 1971 and Cheltenham and Swindon in 1981 compared with the position in 1984. In 1971 Thanet and Lancaster had almost identical local aggregate employment profiles – both had roughly 11,000 in manufacturing, 22,000–23,000 in services and 41,000–43,000 in their local labour market as a whole. But whereas Lancaster sustained its 1971 level of total employment, Thanet's declined by 24 per cent. Moreover, while Lancaster's service sector grew by 17 per cent, that of Thanet, against the national trend, declined by 9 per cent. Finally Thanet, although located in the South, close to London, and having like Lancaster a substantial resort function, lost a greater share of its manufacturing employment, 45 per cent against 36 per cent. These two localities, while both having higher than average service sectors and lower than average manufacturing sectors have become significantly less alike than they were in 1971.

The same can be said of Cheltenham and Swindon in 1984 by comparison with 1981. In the latter year they had total labour markets of similar size and comparable manufacturing and services submarkets. But since 1981 Cheltenham has gone against the national trend and lost 6 per cent of its service jobs which, along with the loss of an identical percentage of manufacturing employment has resulted in a decline of 4000 jobs overall. Swindon, by contrast, has grown in both manufacturing and services employment with the result that its local labour market was in 1984 32 per cent larger than Cheltenham's. Hence, local specificity appears to be returning to some local labour market structures as different labour market sectors interact unevenly with the constraints and enablements of the national and international context.

Statistics such as these, while interesting and informative, do not tell us much about the changing social composition of the UK urban

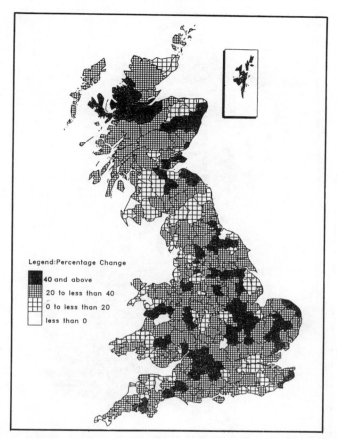

Figure 1.8 Percentage change 1971–1981 in SEG (number of households): employers, managers, farmers, professional and intermediate non-manual. Source: OPCS.

and regional system or of the seven study localities. Furthermore, those statistics that are available for this purpose come from the decennial census, the last of which referred to the position in 1981, so the implications for social structure of employment shifts after that date have to remain unexplored. It will be noted that Figure 1.8 reveals the largest concentrations of growth in the managerial, professional and other white collar socio-economic groups to have occurred in the outer reaches of the metropolitan south of England, in predominantly rural and small or medium-sized localities. Three main geographical areas south of the Severn–Wash line showed

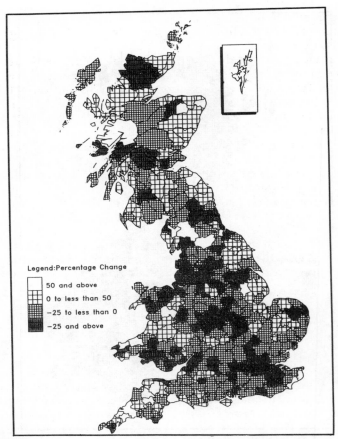

Figure 1.9 Percentage change 1971–1981 in SEG (number of households): foremen, supervisors-manual, skilled manual workers in manufacturing. Source: OPCS and GROS.

an increase in what has been termed the 'service' class of professionals, managers and support staff (Abercrombie and Urry, 1983) of more than 40 per cent between 1971 and 1981. They were, first, the Cheltenham–Swindon–Newbury and neighbouring localities, mostly within an hour or so's travelling time of London, Birmingham and Bristol. An outlier of this growth area could be discerned near the line of the M5 motorway into Devon. Second, there was a large block of service class growth in the area from Cambridge and Bedford to Lincoln, similar in character to the settlement scale and pattern in the first growth area, and comparably centred along good communication links to London and East Midlands cities. The third area was in East Anglia from Colchester to Norwich, including part of the Norfolk coast.

Elsewhere in Britain the concentrations of service class growth were much smaller in geographical extent and mainly situated on the fringes of larger urban concentrations. The 1970s and early 1980s were clearly characterized by a notable and widespread decentralization of higher status households beyond the earlier-established commuting sheds of the cities. But it should also be recalled that the areas of service class growth overlap considerably with those of producer services employment growth. That is, both residential and workplace growth were taking place in these areas. So it is not simply a case of city commuters filling up environmentally pleasing and relatively underpriced residential locations, but rather a mixture of that and some non-metropolitan employment gain in service occupations hitherto found more commonly in cities. The same overlap occurred on a much smaller scale in the northern localities registering service class growth above 40 per cent, reinforcing the argument that this represented a shift in the labour market as well as the housing market dimension of the urban and regional system.

In general terms the spatial-occupational pattern of job loss for households from 1971 to 1981 could be said to be complementary to that of the job gain just discussed. The gains in, for example, supervisory and skilled grades in manufacturing were far less extensive, and mostly found in rural areas, while the losses in skilled occupations were greatest in and around the largest urban labour markets. This could be predicted from the large loss of employment in manufacturing in cities discussed earlier. What was less predictable was the widespread loss of skilled and supervisory manufacturing jobs, admittedly at a lower rate, in the semi-rural fringe localities of the conurbations, as shown in Figure 1.9.

Occupational shift in the seven study localities varied markedly around the Great Britain mean, though the general directionality of the change profiles for particular socio-economic categories tended to be consistent with national change during the 1971–1981 period. Taken as a group the study localities showed marginally less extreme occupational shifts than Great Britain as a whole. There was a lower increase in the service class (employers, managers, professionals and intermediates) and a lower decline in the blue-collar categories of supervisors, skilled and unskilled workers. But at the level of individual localities there were quite dramatic variations, especially in outer Liverpool which lost between a fifth and a quarter of all except its semi-skilled workforce, but showed an above-average rise in self-employment. Middlesbrough, though, which had the highest

Table 1.7 *Change in economically active population by socio-economic group, 1971–1981*

Socio-economic group	GB (%)	SW Birmingham 1971	1981	(%)	Cheltenham 1971	1981	(%)	Swindon 1971	1981	(%)
Employers and managers	+23.6	7530	8560	(+13.7)	6460	8340	(+29.1)	8360	13240	(+58.4)
Professionals and intermediates	+26.3	11360	13920	(+22.5)	8590	10660	(+24.1)	10080	13650	(+35.4)
Supervisory and skilled	−11.0	28050	23990	(−14.5)	10630	9850	(−7.3)	23930	21160	(−11.6)
Semi-skilled	+1.0	48780	47610	(−2.4)	25240	25440	(+0.8)	40890	44310	(+8.4)
Unskilled	−14.6	6690	6920	(+3.4)	3510	3210	(−8.5)	7290	6950	(−4.7)
Own account	+8.5	2610	3030	(+16.1)	2250	2770	(+23.1)	4070	5020	(+23.3)
Unemployed	+92.6	4870	11060	(+127.1)	2200	3250	(+47.7)	3120	9100	(+191.7)

Socio-economic group	Lancaster 1971	1981	(%)	Outer Liverpool 1971	1981	(%)	Thanet 1971	1981	(%)	Middlesbrough 1971	1981	(%)
Employers and managers	7070	7240	(+2.4)	6820	5350	(−21.6)	7260	7610	(+4.8)	10800	13060	(+20.9)
Professionals and intermediates	7940	9470	(+19.3)	9720	7710	(−20.7)	6480	7170	(+10.6)	15180	19470	(+28.3)
Supervisory and skilled	12340	11140	(−9.7)	30720	22480	(−26.8)	11590	1050	(−9.4)	41190	39980	(−2.9)
Semi-skilled	22010	22790	(+3.5)	53770	45770	(−14.9)	22260	22230	(−0.1)	56850	59360	(+4.4)
Unskilled	4700	4200	(−10.6)	14920	10770	(−27.8)	3430	2910	(−15.2)	17970	16150	(−10.1)
Own account	4000	4060	(+1.5)	1770	2010	(+13.6)	3330	3950	(+18.6)	3670	3360	(−8.4)
Unemployed	3310	5940	(+79.5)	12780	22020	(+72.3)	3760	5270	(+40.2)	10470	24690	(+135.8)

Source: Population Census Small Area Statistics (best fit of 1971 and 1981 ward and district data for 1984 travel-to-work areas)

increase in unemployment, performed reasonably close to the mean except for self-employment. Reading the statistics from Table 1.6 showing substantial job loss in manufacturing with its near-average occupational shift from 1971–1981 suggests a higher general growth of employment in services in that period, and a rapid rate of local labour market recomposition. This explanation, involving rates of exit from and entry to the labour market by distinctive social and gender categories would be entirely consistent with the propositions of restructuring theory.

Swindon's large loss of manufacturing employment in the 1970s shows up in the dramatic increase in unemployment registered then. But Swindon is fortunate in terms of location – relatively close to London – in addition to which it continues to benefit from Expanded Town status and the advantage of a high degree of 'local chauvinism' within the political system, as a consequence of which Swindon is well-known as a welcoming location for new domestic and inward investment. Thus, as well as achieving a doubling of service employment between 1971 and 1981, Swindon has been able to outperform the British mean in all occupational categories except supervisory and skilled grades. It is worth noting, too, that Swindon's 58.4 per cent expansion in the employer and manager occupational strata included a female growth component that was almost double the national rate (the increase was 54 per cent in the case of single, and 123 per cent in the case of married women).

Cheltenham's changing occupational structure has also involved a reinforcing of the service class component of the local labour market, but without the rather high loss of supervisory and skilled blue-collar grades observable nationally and in most other study localities. Historically, Cheltenham has kept a foot firmly in both major employment camps. Its early involvement in aeronautical engineering has been maintained by virtue of continuing state expenditure upon military procurement of locally produced aerospace equipment. Similarly, massive state investment in equipment and personnel at GCHQ to undertake signals interception and interpretation for military purposes has bolstered the local economy tremendously. And Cheltenham's history as a spa resort catering for the upper class left it with a privatistic social infrastructure that makes it a continuing magnet for firms and personnel, especially in producer services, committed to the 'enterprise culture'.

Lancaster also developed a broader base in services as manufacturing employment entered early decline. However, in the 1971–1981 period at least, public rather than private services appear to have led this expansion. Thus professional and intermediate service occupations had a far higher growth rate than employers and managers during the 1971 to 1981 period. However, it is worth noting that after 1981 Lancaster had much higher than average employment growth in producer services, especially for part-time female and full-time male employees. Without exaggerating this development, which still leaves Lancaster well behind Swindon and Cheltenham in that sector, it is likely that a combination of income-multiplier effects from earlier public sector investments such as the university and health facilities, a not insignificant element of locally based initiative, and a fortunate location in respect of both high quality environment and communications are transforming Lancaster. From being an essentially Victorian Lancashire industrial town it has become a sub-regional services centre underpinned with the social and political characteristics of a university town, including a range of issue-based social movements. It is also, despite its embourgeoisement, becoming more of a Labour-supporting locality than hitherto. Morecambe, part of the same local government administrative unit, retains the Conservatism typical of its resort tradition but in other respects is suffering from the loss of demand for its tourist attractions. Hotels and guest houses now accommodate temporary construction workers and DHSS claimants more than holidaymakers.

The same fate has, to some extent, befallen Thanet, created with local government reform in 1974 from the formerly independent Margate, Ramsgate and Broadstairs in Kent. The internationalization of British tourism has meant the decline of the traditional seaside holiday. For tourist-dedicated places such as those comprising Thanet there has never been significant industrial employment, and much of what there was has been lost in the recessions of the 1970s and 1980s. However, such has been the crisis faced by local hoteliers that the municipality has become more active than before in seeking to revive the local economy. Political conflict has grown within the ruling group as different interests pursuing fluctuating objectives coalesce, then diverge, often on questions of local loyalty.

The experience of high long-term, unemployment in the three northern urban-industrial localities – Longbridge-Bournville in south-west Birmingham, Kirkby, Speke and Halewood in outer Liverpool, and Middlesbrough – has been associated with divergent socio-political developments. Birmingham had been one of Britain's most successful cities when the motor vehicle and engineering industries were booming. As a consequence workers formed their main associational links in the workplace rather than in the locality. When redundancies were announced workers and their families tended, therefore, to be isolated since even sociability was likely to have centred upon work-based clubs and associations. Hence there has been little formal or informal local response to unemployment outside the factory. The City Council survives as the main political avenue, a role that it shared until 1986 with the West Midlands County Council, one of the most innovative local authorities in the country until abolished by the Thatcher government.

The city of Liverpool has also been a source of tremendous oppositional politics between local and national state apparatuses. The outer Liverpool suburbs were involved to varying degrees, although Knowsley Borough – which contains the largest settlement, Kirkby – has been less directly oppositional in its stance towards central government's cost-cutting policies. A longer experience of high unemployment than, say, south-west Birmingham and a tradition of community associations, some of which are affiliated to the church, has meant that residents of these public sector suburbs built in classically Fordist style to house workers in mass production industries in the 1950s and 1960s, have developed better coping mechanisms and survival strategies than elsewhere. However, these tend to look inward to meet social needs rather than outward to seek private investment. A capacity for mobilizing local loyalty and political influence was an important component in the rise of Middlesbrough as an approximation of the 1960s heavy industry growth pole. The north-east of England, with Middlesbrough an active element, was a good example of a territorial coalition in which diverse interests were obscured for the benefit of the region and its localities. The north-east still retains that political character. However, economic recession has dealt serious blows to Middlesbrough and its organized working class. The trade union movement was the source of much of the political energy and initiative within the local system and, as Middlesbrough's local social structure falls more into line with the services-dominated national profile, that tradition has yet to be rebuilt to its former strength.

These local profiles, delineated in greater detail, form the content of the chapters which follow. Each has clearly experienced substantial change to its local labour market and the social structure which, in part, is formed by such changes. Some of these localities are well placed to take advantage of national and global changes in the processes of spatial development, others are less so. The global perspective reveals clear signs that the dominant mode of organizing economic afairs at the level of the firm or corporation involves a reduction in the emphasis placed on mass production and economies of scale, and a form of restructuring in which flexibility is revalued. This principle is deeply embedded in the organization of the banking, finance and insurance industries, the producer services which, in the UK context, are crucially dependent on London. London's impact on the UK urban and regional system has been and continues to be fundamental. While giving, as it were, opportunity to southern towns with sufficient local initiative to take advantage like Swindon (but unlike Thanet) of private sector growth, London's governmental arm takes away the means for developing local initiative in the northern cities where public expenditure is a much more important component of local economic health. The case for a reconsideration of this manifest infringement of social justice, and of the starkly uneven consequences for those divided by geography and class from prosperity, will be outlined in the concluding chapter of this book.

References

Abercrombie, N. and Urry, J. (1983) *Capital, Labour and the Middle Class*, London: Allen and Unwin.

Atkinson, J. (1984) *Manning for Uncertainty*, Institute of Manpower Studies Report, Sussex University.

Austrin, T. and Beynon, H. (1979) 'Global outpost: the working class experience of big business in the north-east of England, 1964–1979'. Department of Sociology discussion paper, Durham University.

Begg, I., Moore B. and Rhodes, J. (1987) 'Economic and social change in urban Britain and the inner cities', in V. Hausner (ed.), *Critical Issues in Urban Economic Development*, vol. 1, Oxford: Clarendon Press.

Champion, A. (1987) 'London's population revival' (mimeo).

Cohen, R. (1987) 'Policing the frontiers: the state and the migrant in the international division of labour', in J. Henderson and M. Castells (eds.), *Global Restructuring and Territorial Development*, London: Sage.

Cooke, P. (ed.) (1986a) *Global Restructuring, Local Response*, London: ESRC.

Cooke, P. (1987) 'Britain's new spatial paradigm: technology, locality and society in transition', *Environment and Planning, A*, 19, pp. 1289–1301.

Cooke, P. and Morgan, K. (1985) 'Flexibility and the new restructuring: locality and industry in the 1980s', *Papers in Planning Research 94*, Cardiff: University of Wales.

Cooke, P. and Morgan, K. (1986) 'Britain goes on a bender', *Times Higher Education Supplement*, 699, March.

Edgell, S. and Duke, V. (1986) 'Radicalism, radicalization and recession: Britain in the 1980s', *British Journal of Sociology*, 37, pp. 479–512.

Fröbel, F., Heinrichs, J. and Kreye, O. (1980) *The New International Division of Labour*, Cambridge: Cambridge University Press.

Giddens, A. (1984) *The Constitution of Society*, Cambridge: Polity Press.

Green, F. (1986) 'Some macroeconomic omens for Reagan and Thatcher', *Capital and Class*, 30, pp. 17–30.

Henderson, J. and Castells, M. (1987) Introduction to J. Henderson and M. Castells (eds.), *Global Restructuring and Territorial Development*, London: Sage.

Hill, R. (1987) 'Global factory and company town: the changing division of labour in the international automobile industry', in J. Henderson and

M. Castells (eds.), *Global Restructuring and Territorial Development*, London: Sage.

Hobsbawm, E. (1968) *Industry and Empire*, Harmondsworth: Penguin Books.

Johnston, R., and Pattie, C. (1987) 'A dividing nation? An initial exploration of the changing geography of Great Britain, 1979–1987', *Environment and Planning A*, 19, pp. 1001–1013.

Lipietz, A., (1982) 'Towards global Fordism?' *New Left Review*, 132, pp. 33–47.

Marshall, R. (1986) 'Working smarter', in D. Obey and P. Sarbanes (eds.), *The Changing American Economy*, Oxford: Blackwell

Marshall, T. (1977) *Class, Citizenship and Social Development*, Chicago: Chicago University Press.

Marquand, J. (1983) 'The changing distribution of service employment', in J. Goddard and A. Champion (eds.), *The Urban and Regional Transformation of Britain*, London: Methuen.

Massey, D. (1984) *Spatial Divisions of Labour: Social Structure and the Geography of Production*, London: Macmillan.

Massey, D. (1987) 'Heartlands of defeat', *Marxism Today*, July, pp. 18–23.

Massey, D. and Meegan, R., (1982) *The Anatomy of Job Loss*, London: Methuen.

Mills, D. (1973) 'Suburban and exurban growth', in Open University (ed.), *The Spread of Cities*, Milton Keynes: Open University.

Morgan, K. (1986) 'The spectre of "Two Nations" in contemporary Britain', *Catalyst*, 2, pp. 11–18.

Nelson, K. (1986) 'Labour demand, labour supply and the suburbanization of low-wage office work', in A. Scott and M. Storper (eds.), *Production, Work and Territory*, London: Allen and Unwin.

Noyelle, T. (1989) 'Services and the world economy: towards a new international division of labour', in P. Cooke and N. Thrift (eds), *Captive Britain*, Cambridge: Cambridge University Press.

OECD (1986) *Flexibility in the Labour Market: the Current Debate*, Paris: Organization for Economic Cooperation and Development.

Piore, M. and Sabel, C. (1984) *The Second Industrial Divide*, New York: Basic Books.

Portes, A. and Sassen-Koob, S. (1987) 'Making it underground: comparative material on the informal sector in Western market economies', *American Journal of Sociology*, 93, pp. 30–61.

Rajan, A. (1987) 'Services – the second industrial revolution?' *Institute of Manpower Studies Report*, Sussex University.

Sassen-Koob, S. (1984) 'The new labour demand in global cities', in M. Smith (ed.), *Cities in Transformation*, London: Sage.

Sassen-Koob, S. (1987) 'Issues of core and periphery: labour migration and global restructuring', in J. Henderson and M. Castells (eds.), *Global Restructuring and Territorial Development*, London: Sage.

Savage, M., Barlow, J., Duncan, S. and Saunders, P. (1987) '"Locality research": the Sussex programme on economic restructuring, social change and the locality', *The Quarterly Journal of Social Affairs*, 3, pp. 27–51.

Schumpeter, J. (1943) *Capitalism, Socialism and Democracy*, London: Allen & Unwin.

Scott, A. (1982) 'Locational patterns and dynamics of industrial activity in the modern metropolis', *Urban Studies*, 19, pp. 111–142.

Simmie, J. (1985) 'The spatial division of labour in London 1978–81', *International Journal of Urban and Regional Research*, 9, pp. 557–570.

Thrift, N. (1987) 'The fixers: the urban and geography of international commercial capital', in J. Henderson and M. Castells (eds.), *Global Restructuring and Territorial Development*, London: Sage.

Thrift, N., Leyshon A. and Daniels, P. (1987) 'Sexy greedy: the new international financial system, the City of London and the south-east of England' (mimeo).

Turner, B. (1986) *Citizenship and Capitalism. The Debate Over Reformism*, London: Allen and Unwin.

Webber, M. (1964) 'The urban place and the non-place urban realm', in M. Webber (ed.), *Explorations into Urban Structure*, Philadelphia: University of Pennsylvania Press.

Westaway, J. (1974) 'The spatial hierarchy of business organizations and its implications for the British urban system', *Regional Studies*, 8, pp. 145–155.

Wood, P. 1987) 'The South-east', in P. Damesick and P. Wood (eds.), *Regional Problems, Problem Regions and Public Policy in the United Kingdom*, Oxford: Clarendon Press.

2

Living in the Fast Lane: Economic and Social Change in Swindon

KEITH BASSETT, MARTIN BODDY, MICHAEL HARLOE & JOHN LOVERING[1]

Introduction

In 1841, with a population of only 2500, Swindon was selected by the Great Western Railway as the site for the rail engineering works which dominated the town up to the last war. Wartime industrial relocation and the postwar boom in consumer goods industries, electrical engineering and the car industry then brought a wave of new employers and a major influx of population. Collapse of manufacturing employment in the 1970s was more than offset by rapid expansion of financial and business services, distribution and new manufacturing sectors including plastics, electronics and pharmaceuticals. The town's workforce has virtually doubled since the early 1950s and its population expanded from 91,000 in 1951 to 151,000 by 1981. Until the last war it had all the characteristics of a northern industrial town, but one in the middle of rural Wiltshire. By the late 1980s it was a key growth centre in the booming M4 corridor.

This account looks first at successive phases of expansion which have reworked the town's economic and employment structure. It identifies the key role of new employers in job growth, the increased 'internationalization' of the 'local' economy, the changing gender composition of the workforce, and shifts in terms of corporate

[1] Keith Bassett works in the Department of Geography, University of Bristol, Martin Boddy and John Lovering at the School for Advanced Urban Studies, University of Bristol, and Michael Harloe in the Department of Sociology at Essex University. We would like to thank Gill Court and Jane Wills, research students at SAUS, for their contribution to the work on which this chapter draws. The main body of research was supported by the Economic and Social Research Council's 'Changing Urban and Regional System' initiative, grant DO 4250015. Some supporting information is drawn from parallel work carried out under the ESRC's 'Social Change and Economic Life' (SCEL) initiative, grant G 13250014 – specific reference is made to information drawn from this source, which drew on extensive surveys of individuals and employers.

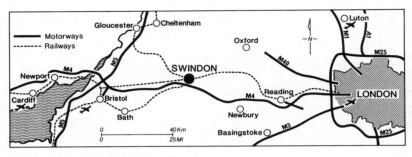

Figure 2.1 The Swindon locality.

strategy and forms of work. It evaluates the processes underlying these changes, focusing in particular on the role of the local authority and the capacity for local 'pro-activity'. Finally, it looks at the complex interrelationship between economic restructuring and changes in class structure, politics and culture.

Economic change

Swindon's selection as the site for the rail engineering works reflected its strategic location on the GWR network – a junction of two lines where the gradient required a change of locomotives and canals supplied both coal and water (Peck, 1983, pp. 8–10). Within a few years, a major, integrated industrial complex had been created. The Works expanded dramatically, producing locomotives, rolling stock and other railway equipment. By 1892 over 10,000 men were employed on one of the largest industrial sites in Europe and the town's population grew from around 7000 in 1861 to 51,000 by 1911, sharply differentiated from the surrounding area in economic, social and cultural terms (Table 2.1).

Between the wars expansion of road transport, and the collapse of the Welsh coal trade which had been a major source of traffic, brought an end to the era of guaranteed prosperity for the GWR, the railworks and the town. Employment fell from a peak of around 15,000 in the mid–1920s. Decline was, however, more gradual than in other localities dependent on heavy industry, the GWR benefiting from the growth of holiday traffic to the south-west and commuting into London. Apart from the railworks however, there were few significant employers before the Second War – a number of clothing companies, Wills cigarettes, and the Garrard record player company which had moved out from London.

Table 2.1 *Population growth, Swindon, 1841–1981*

	1841	1861	1881	1901	1921	1931	1951	1961	1971	1981	1991 est.
Swindon	2,495	6,865	19,904	45,006	50,751	77,873	–	–	–	–	–
Thamesdown	–	–	–	59,285	65,890	–	90,570	119,451	139,351	150,746	

Source: Population Census. Figures for current Thamesdown Borough area are reproduced from Thamesdown Borough Council (1984); figures for Swindon from Harloe (1975)

Plate 2.1 Swindon and its railyards in the 1950s.

The postwar boom

With wartime manufacturing growth Swindon became a major military industrial complex. Production at the GWR works switched to military needs, expansion away from high bomb-risk areas brought major new establishments including Short Brothers, Armstrong-Whitworth and Plessey, and in-migration of war workers boosted the town's population by 16,000 between 1939 and 1941. Military bases were also established close to the town, and several remain as major employers. Wartime production had major impacts in many localities – Cheltenham and outer Liverpool are other examples. In some cases it left little trace. In Swindon's case, however, wartime production was the foundation for postwar modernization and diversification and represented a first phase of planned expansion.

Growth of existing employers was important initially. Rearmament benefited firms like Plessey and Vickers, and reconstruction boosted demand at the rail works, nationalized in 1948. Firms such as Garrard and Plessey also benefited from protected domestic markets, and the boom in UK manufacturing exports. From the mid-1950s however, new employers contributed increasingly to employment growth. The Pressed Steel car body plant was particularly significant. This was established in 1955 after the government denied the company permission to increase capacity at Cowley. Enthusiastically supported by the borough council, the new plant was a major factor allowing the town to capture London's overspill and establish its credibility as a growth point. By 1961 the company employed over 4000, including around 1000 ex-Londoners and 600 skilled men from the railworks.

British Railway's 'modernization plan' largely staved off decline in rail engineering employment, leaving the number of jobs at around 10,000 in the early sixties – only 2000 down on 1950. Overall, therefore, the 1950s was a period of particularly rapid expansion. The town's population increased by around 3000 per year throughout the decade and employment increased by nearly 50 per cent in the ten years from 1952 to 1961. This reflected, above all, the boom in manufacturing, but employment remained concentrated in the engineering and vehicles sectors.

Strong growth was maintained throughout the 1960s, with employment up by 9000 (15 per cent) over the decade to 1971. Within a continued in-migration of new employers, the American high tech plastics company Raychem and the pharmaceutical company Roussel represented a new wave of expanding manufacturing industry. Manufacturing employment did not peak in absolute terms until 1970, four years after decline set in nationally, and significant job loss in the 1960s was confined to the railworks. There employment was virtually halved between 1965 and 1967. This reflected 'rationalization' of the national rail network under Beeching, decreased demand for maintenance and replacement with modern locomotives and stock.

The 1970s

The 1970s saw a fundamental shift in the town's economic and employment base. Heavy manufacturing job loss coincided with in-migration of new companies and marked growth of service employment. Unemployment climbed rapidly, as elsewhere, and heavy manufacturing job loss temporarily pushed the rate above national levels in 1974 (Figure 2.2). But, at a time when many local

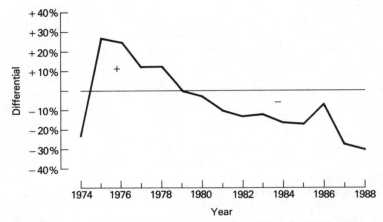

Figure 2.2 Unemployment, Swindon and Great Britain, 1974–1987.
The diagram shows the unemployment rate in Swindon travel-to-work area as a percentage of the rate for Great Britain. Areas above the central axis indicate that the rate in Swindon was higher than the national rate and areas below the central axis that it was lower. Rates shown are for June each year except for the final figure which is for December 1987.
Source: Department of Employment

economies faltered and went into decline, Swindon successfully diversified away from manufacturing.[2] With the onset of recession nationally, growth was slower than in the 1960s, but total employment still expanded by 10 per cent over the decade compared with a 2 per cent drop nationally. There were, however, far-reaching economic and social implications. A third of the town's manufacturing employment was lost over the decade to 1981, service employment more than doubled, and there was a major increase in female employment (see below).

Major job losses were concentrated in electrical engineering, vehicles, metals and other engineering (Figure 2.3). Rail engineering employment was relatively stable after major losses over the previous decade. Plessey, among the town's largest employers, disposed of all but one of its Swindon companies, including Garrard: a heavy loss of

[2] Overall and sectoral employment figures are from the Department of Employment, Annual Census of Employment. Figures and information relating to individual companies derive from detailed case studies of a panel of local employers carried out in 1986 and 1987. All statistical information relates to the Swindon travel-to-work area, unless otherwise noted. The TTWA extends beyond the area of Thamesdown Borough Council, although the latter includes the major employment centre. The local authority boundary was extended in 1984 to include the Western Expansion and the name changed from Swindon to Thamesdown.

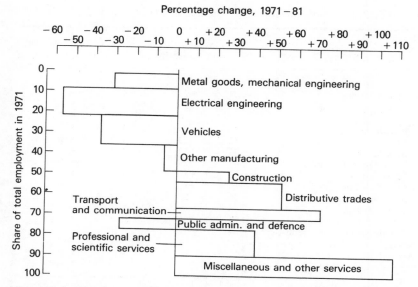

Figure 2.3 Employment change, Swindon, 1971–1981 and 1981–1984.

1 The horizontal axis shows the proportion of total employment in each sector in the base year. The vertical axis shows percentage change in each sector over the period. The area of each column is therefore proportional to change in absolute terms.

2 Figures for 1971–81 are based on the 1968 Standard Industrial Classification and those for 1981–84 on the 1980 Classification. The two are not, therefore, directly comparable. Public utilities are included with 'other manufacturing' for both periods to increase comparability. Financial and business services are included with 'miscellaneous services' for 1971–81 for presentational purposes – they recorded an increase of 420% from a base of 1.6% in 1971.

Source: Department of Employment, Annual Census of Employment, NOMIS

jobs. Garrard itself, employing 4000 in 1973, was hit hard by Japanese competition and finally closed in 1979. The domestic car industry ran into increasing difficulties in both export and domestic markets. Employment at the Swindon plant fell from around 5000 in 1979 to 3400 by 1981. New technology and major productivity gains also contributed to job loss although the plant, by now part of the Rover Group, remained a major local employer with 1986 employment of around 3000.

Major job gains through the 1970s were concentrated in financial services, distribution, professional and scientific services including education and medicine, and a range of other services. Modest net gains in some manufacturing sectors up to 1978 turned to losses virtually across the board thereafter. In-migration of new employers

Table 2.2 *Employment change, Swindon and Great Britain, 1971–84*

Category	% Change 1971–81		% Change 1981–84	
	Swindon	Great Britain	Swindon	Great Britain
All	+10	− 2	+ 7	− 3
Manufacturing	−34	−24	+ 3	−11
Service	+52	+15	+ 9	+ 5
Male	− 1	− 6	+ 8	− 3
Female	+27	+ 3	+ 5	+ 2
Male manufacturing	−31	−23	+ 1	−12
Male service	+44	+15	+15	+ 4
Female manufacturing	−44	−29	+16	− 9
Female service	+60	+15	+ 4	+ 5
Male full-time	− 3	−11	+10	− 4
Male part-time	−50	+20	−14	+10
Female full-time	+13	− 4	+15	+ 1
Female part-time	+51	+35	− 7	+ 3
Female manufacturing full-time	−34		+18	− 8
Female manufacturing part-time	−68		+ 6	−15
Female service full-time			+16	+ 4
Female service part time			− 8	+ 6

1   1968 SIC III–XIX
2   1968 SIC XXII–XXVII
3   1980 SIC 4–25
4   1980 SIC 27–36
   Source: Annual Census of Employment, Department of Employment

accelerated through the 1970s, with the emphasis shifting towards office-based employment and distribution. The Post Office established a national supplies centre, W. H. Smith set up a warehouse and distribution facility for its national retail chain, Burmah Oil established a purpose-built administrative headquarters, and Allied Dunbar set up its first Swindon office (see Table 2.3).

The 1980s

Nationally, there was a sharp downturn in the economy in the early 1980s with heavy loss of manufacturing employment. This resulted in rising unemployment locally. Swindon, however, survived relatively well in terms of the national picture (Figure 2.2). Employment grew by seven per cent in the three years 1981–84, compared with a one per cent drop nationally.[3] There was strong growth in service employ-

[3] A survey of employers in 1986 suggested an overall increase in employment locally of around 14 per cent between 1981 and 1986. The survey was carried out by Michael White of the Policy Studies Institute for the ESRC SCEL initiative. It comprised a telephone survey of 240 local employers in the private and public sectors, which accounted for over 50 per cent of total employment in the Swindon travel to work area.

Table 2.3 *Incoming companies*

Company in Swindon	Established	Activity	Current employment
Allied Dunbar	1971	Financial services	1,760
Anchor Foods	1979	Packaging and national distribution of butter and other food products	360
Book Club Associates	1968	Book club administration, warehouse and distribution	860
BR SW Region	1984	Regional HQ	790
Burmah Castrol	1973	UK Lubricants and Fuels HQ and Group HQ	800
Intel Corporation	1979	Semiconductor marketing, sales, distribution and design	360
Nationwide Anglia B S	1975	Computer and administrative centre	450
NEM	1980	Financial services	230
Post Office	1973	National supplies, engineering and research division	550
Raychem	1966	Specialist plastics manufacture	1,300
Renault UK	1983	National parts distribution and training centre	187
Rousell Laboratories	1970	Manufacture of pharmaceutical products	610
R. P. Scherer	1982	Manufacturer of pharmaceutical products	260
SERC/NERC	1978	Government research council administration	650
W. H. Smith	1966	National distribution centre and warehouse, Retail Division HQ	1,400

The table lists a selection of the larger employers which have relocated to, or set up in Swindon since the mid-1960s together with the most recent information on total employment. This is not a complete list of new employers.

ment – up by nearly nine per cent – and manufacturing employment actually expanded by three per cent, compared with an eleven per cent drop nationally (Table 2.2).

Local job loss was again concentrated in the longer established engineering and metal goods sectors, including rail engineering (Figure 2.3).[4] But there was significant growth (over 2000 jobs) in chemical and non-metal products, and in electrical and electronic

[4] Figures for employment change rely on results from the 1984 Annual Census of Employment at Travel to Work Area level, which, prior to checks on accuracy, must be taken as preliminary.

engineering: all these sectors recorded strong growth locally, compared with static or declining employment nationally. Service employment growth was concentrated in retailing, up by twenty-one per cent compared with static employment nationally, and in insurance, finance and business services, up by twenty-nine per cent, compared with seventeen per cent nationally. Together they added around 2800 jobs.

After 1984, unemployment remained significantly below national levels. Growth continued in financial services: Allied Dunbar, for example, expanded employment by nearly 500 between 1984 and 1988. Honda set up a UK pre-sales service centre on the old Vickers airport and was also establishing a fully automated engine plant on the site. These decisions were probably influenced by its joint production deal with Austin Rover and, as noted in Chapter 7, have raised speculation that car assembly might follow.[5]

In 1984 major redundancies were announced at the railworks followed by closure in March 1986 and the loss of the remaining 2400 jobs. British Rail once again blamed the switch of freight to the roads, and decreased demand for maintenance and replacement. However, rationalization was influenced by the government's requirement that rolling stock be procured wherever possible by competitive tendering, and the unions argued that the main rationale was to make room for the private sector and prepare rail engineering itself for privatization. The flow of new establishments continued but these were primarily smaller companies rather than major relocations – W. H. Smith's retail head office employing around 150 was among the more significant moves.

New employers and job growth

Job loss in manufacturing has been common to most localities. It is the scale of job gain in services and to some extent in manufacturing, however, that distinguishes Swindon. The key to continued expansion has been the attraction and subsequent expansion of new employers. Nearly 240 new employers who set up in the town between 1957 and 1982 still operated locally by the end of that period: over 100 in manufacturing, 80 in distribution and 40 in services, including several which are now major employers (Thamesdown Borough Council, 1984a). Simple relocation of existing establishments was important in some cases, combining in-migration of existing staff with significant local recruitment – W. H. Smith and Burmah Oil are examples. In other cases, initial employment was

[5] The outcome and possible implications of the British Aerospace bid for the Rover Group in March 1988 were unclear at the time of writing.

small, and it was subsequent growth which generated significant employment.[6] Allied Dunbar began by employing around 100, but expanded rapidly to become one of the major local employers with a prominent physical presence in the town centre and a workforce of nearly 1800 by 1987. Book Club Associates set up under the wing of W. H. Smith now employs nearly 900 and Raychem, set up in 1966 with 17 employees now totals 1300. Many new arrivals were relocating specifically in order to expand while companies like Allied Dunbar, Book Club Associates and Raychem were able to capitalize on rapidly expanding markets.[7]

Internationalization

Increased competition in overseas markets, import penetration and inward investment have had marked and varied impacts in different localities. In Swindon's case collapse of export markets, removal of tariff barriers and increased penetration of UK markets, together with lack of investment, led to major rationalization and manufacturing job loss from the early 1970s, in firms like Plessey, Garrard and Austin Rover. Plessey refocused its activities on semi-conductor design and manufacture, employing around 750 locally in 1987, and operating successfully in a highly competitive international market. The Austin Rover plant expanded production for Saab, Volvo and Honda in order to decrease dependence on its parent company.

Inward investment by foreign-owned, particularly US companies (mainly in electronics, plastics and pharmaceuticals), first evident in the 1960s, has been increasingly significant. US multinationals increased their share of manufacturing employment in the town from 15 per cent in 1979 to around 27 per cent by 1985 (firms employing over 100) and total employment in US–owned companies was around 6500. Swindon was thus a major beneficiary of the increasing internationalization of the UK economy, attracting primarily US investment. Other major employers in recent rounds of investment, such as Allied Dunbar or W. H. Smith, are oriented primarily to UK national markets. A smaller number of companies, such as Raychem and thus US–owned semi–conductor company Intel, have a more European orientation.

[6] In the 1986 SCEL Employer Survey, 48 per cent of employers reported an increase in employment over the previous five years compared with only 17 per cent who reported a drop in employment. Fifty-eight per cent of larger private sector employers (more than twenty employees) reported an increase in jobs.

[7] In the 1986 SCEL Employer Survey (note 3), 70 per cent of larger private sector employers reported increased demand for products or services over the previous five years, and only 8 per cent a decrease. Increased demand was reported to be the overwhelming factor likely to lead to future employment growth.

The changing gender division of labour

Increased female participation in paid employment, particularly part-time, has been a marked feature of labour force recomposition nationally. The scale and nature of gender recomposition has varied, however, across different localities. Swindon was until recently characterized by particularly low rates of female participation – 25 per cent locally in 1931 compared with 34 per cent nationally.[8] This reflected the dominance of male employment in rail engineering, which significant employers of women like Wills, Garrards, Plessey and the clothing companies did little to offset. Female employment in manufacturing expanded in the postwar boom, but growth in Pressed Steel and other companies largely maintained the dominance of male manufacturing employment. It was not until the 1960s that female participation rapidly caught up with and overtook national levels.

Low female activity rates, therefore, in part reflected lack of opportunity for paid employment due to the town's industrial structure. How far this was perpetuated in household and workplace attitudes which discouraged women from working is hard to determine. There is some evidence, however, of shortages of female employment in the 1960s, when participation rates were still relatively low (Hudson, 1967). Demographic factors played a part: many women in households which migrated to the town in the main expansion phase from the mid-1950s were involved in child rearing when they first came. Availability of female labour increased when they, and later their children, came on the labour market.

Growth in female employment was particularly rapid in the 1970s, expanding by 27 per cent (7500) in the decade to 1981, compared with only 3 per cent nationally (Table 2.2). By 1981, 50 per cent of women had paid employment compared with 46 per cent nationally, reversing the earlier picture,[9] and reflecting the massive increase in service employment noted earlier. Female employment in manufacturing fell sharply, but female service employment grew by 60 per cent, far outstripping national trends. As elsewhere, much of this increase (over three-quarters) represented part-time employment which more than doubled. For many incoming firms, the ready availability of female labour over this period was an important consideration and companies such as Burmah, W. H. Smith, Book Club Associates and Allied Dunbar are all significant employers of women. Female clerical employment in particular increased – from 4000 to nearly 9500 in the decade to 1981. Increased employment in public sector services and retailing were other important growth areas.

---

[8] Figures refer to Swindon Borough.   [9] Figures refer to Thamesdown Borough.

Plate 2.2   The Murray John Tower, Swindon town centre.

Over the 1970s there was, therefore, a major shift in the overall balance of the workforce towards female employment, reflecting in particular growth in female part-time employment. In the early 1980s, however, male employment grew faster than female employment (Table 2.2) and, in marked contrast to the 1970s, part-time female employment actually fell – by seven per cent compared with continued though modest growth nationally. This reflects a trend common to several of the more prosperous, M4 corridor localities. This loss of part-time female employment was concentrated in public administration, defence and medical services, around 2000 down between 1981 and 1984, only partially offset by an increase in part-time employment in retailing. Full-time female employment, on the other hand, grew strongly.

Changing industrial structure, particularly the switch from manufacturing to services, has been the major factor in gender recomposition in the longer term. Employment has grown in sectors characterized by a relatively high proportion of female employment. This includes sectors in which a significant proportion of employment is in jobs defined as 'women's work' (horizontal segregation). Employment has also expanded in activities such as financial services in which there is significant female employment but where it is concentrated disproportionately in lower grades (vertical segregation).[10] Examples of 'feminization', in the sense of female inroads into formerly male occupational preserves, are rare. Despite the dynamism of the local economy and rapid employment growth in some sectors, there is little evidence, therefore, that gender divisions at work have been significantly eroded.

Corporate strategies and new forms of work

Economic expansion and diversification have been reflected in major changes in work culture, organization and control. Postwar growth of engineering imported traditional 'Fordist' work practices based on firm hierarchical control, the skilled manual proletariat and large-scale flow-line production processes. These are still reflected in some sections of manufacturing. With continued in-migration, however, and fewer long-established employers than in many localities, forms of work and management-labour relations owe less to earlier rounds of investment and represent a cross-section of more contemporary patterns of corporate strategy and work structure. Incoming employers have imported a variety of pre-existing management practices and work cultures. Relocation and expansion also, however, afford opportunities for social and technical innovation, including the introduction of new technology and work practices.

Rapid decline in manufacturing from the late 1960s hit male, hierarchically structured and heavily unionized employment. Sub-

---

[10] For discussion of horizontal and vertical segregation, see Hakim (1979).

sequent expansion has greatly increased the number and diversity of employers, forms of work and work culture. Firstly, office-based clerical labour has become increasingly important. Hierarchically structured, this is typically characterized by consensual management styles and an absence of unions. It includes 'career' grades characterized by clear promotional paths, professional identification and a high degree of vertical segregation along gender lines. There has also, however, been a major expansion of routinized clerical employment, primarily female, such that sections at least of some of the larger office-based employers are effectively information processing 'clerical factories' but without the traditions of shopfloor organization. Work pressures can be keen: 'you have to live and die for work, you have to do overtime and to be there when they want you' (a), and 'although the money's good, they expect you to work like slaves . . . they push you to the limit, it's work, work, work' (b).[11]

Secondly, there has been a major growth of relatively low-paid employment in retailing, cleaning, catering and other services. Much of this is female employment and much of it is part-time. In the private sector, it is predominantly not unionized and offers little job security: 'there's no security, there is no such thing. It doesn't matter how loyal you are to your company now . . . you are just one of a number. If we want you this week, we'll have you. If we don't want you next week, we'll get rid of you' (c). The public sector, on the other hand, is characterized by greater job security and union recognition.

Finally, several of the more prominent employers have established new forms of personnel practice, drawing in part on the 'new human relations' approach to personnel practices typical, for example, of American electronics companies. The aim is to combine job satisfaction with consensus, commitment and control. This typically involves single-status employment, performance-related pay and benefit structures, recruitment emphasizing personal characteristics and 'company fit' as much as formal qualification, abbreviated chains of command emphasizing communication and contact rather than formal hierarchy, flexible work practices, and pre-emption of unions. American companies like Intel or Raychem have imported such practices virtually intact, whereas others have combined some features of the model into their particular work culture. In other

---

[11] Quotes here and following are from semi-structured interviews carried out with a panel of households in Swindon TTWA in 1987. The panel was grouped into five categories: professional and managerial workers, female clerical workers, women working part-time, male workers in traditional manufacturing, and unemployed. The selection frame for this panel was the SCEL survey. Letters in brackets refer to the appendix which gives brief details of respondents quoted.

cases, more traditional structures apply to manual grades. Manual workers in W. H. Smith and Book Club Associates, for example, are fully unionized, reflecting the established strength of the print union Sogat prior to the company's move from London – staff are referred to internally as 'union' and 'non-union'. In other cases, manual workers are incorporated in a unified approach. In many newer employers, both in manufacturing and in office-based services, the number of employees receiving benefits such as private health care, financial assistance with housing, company cars and company share schemes has increased significantly. A survey in 1986 indicated that 25 per cent of those in employment aged 20–60 had private health insurance through their employer, compared with 9 per cent of employees nationally.[12]

Employment growth has also been accompanied by qualitative change in the way in which labour is used. In some cases, use of part-time labour has been increased to achieve flexibility, particularly in the volume of labour inputs.[13] As elsewhere, this is common in contract cleaning and much of retailing, for example. It is also used, however, in order processing activities like the book clubs, and one major employer has adopted an explicit strategy of job splitting for a significant proportion of its lower grade clerical labour. A number of newer employers make significant use of temporary labour to meet seasonal and other peaks in demand. Labour is called in as needed from a pool of temporary workers maintained on company books, and permanent employees are generally recruited from these 'temps' as needed. Other companies have developed quite complex shift systems designed in part to meet company needs in terms of labour inputs and to tap into specific sections of the labour market – women working part-time, for example. Again, this includes employers across a range of different sectors such as electronics, retailing and mail order.

There is evidence, therefore, of increased numerical flexibility. There are counter examples, however, of temporary work being replaced by single-status employment in the interests of cooperation

---

[12] Figures for Swindon are from a random survey of one thousand individuals aged twenty to sixty, resident in Swindon TTWA, carried out by Public Attitude Surveys in 1986. This formed part of the ESRC SCEL initiative, Swindon Study, carried out by Martin Boddy and John Lovering, together with Jonathon Gershuny, Michael Rose and Sarah Fielder of Bath University. References to 'the 1986 survey' later in the chapter relate to this work. The national figure is from Social Trends 1987, OPCS.

[13] The 1986 SCEL Employer Survey indicated that 23 per cent of employers had increased their use of part-time employment over the previous five years and 10 per cent had replaced full-time with part-time employment. Increased flexibility in terms of labour inputs was the main motive.

and productivity. And there is much less evidence among local employers of task or function-based flexibility. To the extent that task-based flexibility is found, it is not, moreover, confined to a core of more privileged workers as some theories would suggest (Atkinson, 1984): it is a key feature in some sections of retailing, for example. Despite the contemporary nature of its employment practices, there is little evidence overall that post-Fordist 'flexible acculation' is particularly important locally. And routinzed operations closer to 'Fordist' models are, as indicated, common in parts of the much-expanded service sector. The overall picture, therefore, exhibits increased complexity rather than any clearly emerging trend.

Economic expansion of the role of local policy

Local policy, committed to growth and physical expansion since 1945, has played a key role in Swindon's economic and social transformation. It was, however, the town's location within the changing urban and regional system of the UK, and in particular its location relative to the south-east economic heartland, which enabled it to sustain its trajectory of growth and diversification through a number of distinct phases which built on and reworked what had gone before.[14]

Town expansion, 1945–64

There was concern locally, even before the war, over declining rail employment. Wartime expansion and relocation, however, radically altered the policy context and provided an initial basis for town expansion and diversification. Council policy immediately after the war favoured retention of wartime industry, provision of new trading estates, central area redevelopment, new neighbourhood areas and satellite villages – precisely the pattern of development realized over the next thirty years. National policy was not initially favourable. The wartime Barlow and Abercrombie Reports set the context for decentralization of industry and population to new growth centres. Regional policy, however, sought to direct industry to designated development areas, restricting development elsewhere by the use of Industrial Development Certificates, and Swindon was considered too large and too far from London to form the nucleus of a New Town under the 1946 Act.

Election of the 1951 Conservative government, however, and

---

[14] The origins and progress of town expansion are analysed in detail in Harloe (1975) which covers the period up to the mid-1960s.

concern over the slow pace of decentralization to new towns, brought a change in the national policy context. The 1952 Town Development Act provided for 'overspill' agreements between designated 'expanded towns' and local authorities in the conurbations. Swindon's designation under the Act, with ministerial approval for expansion by around 23,000 to 92,000, was rapidly accomplished, and agreements were secured with Tottenham and with the London County Council (Harloe, 1975).

Board of Trade approval for Pressed Steel in 1955, for which the local council fought hard, and subsequent relaxation of IDC controls, marked the start of new industrial growth. Policy change was reinforced by its coincidence with the boom in UK manufacturing industry. Initially, relocation of manufacturing firms, particularly from London, was an important component. Decentralization was notably encouraged by the shortage and escalating cost of labour and premises in the capital, compared with Swindon. Engineering skills were readily available, together with an expanding supply of semi-skilled and unskilled labour, facilitated by the provision of local authority housing. Proximity to London was also a factor, many firms seeking to minimize the distance moved. Swindon was particularly successful in its use of the 1952 Act. Initial expansion was rapid, with population overspill from London, around 20,000 on some estimates, an important factor. The pace of growth slackened in the mid-1960s, reflecting tighter IDC controls, slower economic growth and competition from the other New Towns closer to London. New employers had been attracted in, but contrary to the council's postwar hopes, only limited diversification away from engineering and vehicles had been achieved.

A strategic place in the south-east, 1964–73

Rapid population growth in the south-east, and pressure for development beyond the London green belt, prompted a series of strategic planning studies, the 1964 South-east Study in particular. In 1965 consultants identified the Newbury–Hungerford area to the east as a possible new city location but this was later rejected in favour of Swindon. Swindon's enthusiastic response to strategic growth proposals, plus its track record of successful expansion, were key factors. This contrasted with opposition from residents and local authorities closer to London. The report proposed Swindon's expansion into a new city with a population of 250,000 by 1981 and 400,000 by 2001.

Population targets were reduced subsequently, but plans, approved by central government in 1968, formed the blueprint for development through to the 1980s. The council set out to realize the possibilities for expansion offered by the regional context in this period, and took an active role in the development process. It bought up land ahead of needs, developing industrial estates, sites for office and commercial development, and private housing. It also invested heavily in social and community facilities. The Western Expansion, with a planned population of around 30,000 based on urban villages and district centres, was a key component. Here the local authority played a leading role as landowner and developer in partnership with the private sector. Central government continued to back development proposals: for example, it granted planning permission in 1978 for a 600–acre residential development when Thamesdown and the developers appealed against Wiltshire County's refusal to grant planning approval.

Decisions at national level on infrastructure development consolidated Swindon's position within the western growth corridor. The choice of Heathrow as Britain's international airport in 1941 was a formative early decision. The routing of the M4 motorway, opened in 1971, past the southern edge of town was crucial, and emerged from protracted negotiations in which Swindon's status as an expanding town was a factor (Hall et al., 1987, p. 165). The M4 gave access to Heathrow and London and tied Swindon in to the expanding national motorway net. This emphasized its growth potential in terms of regional and national distribution and services, realized through the 1970s with the arrival of employers like Anchor Butter, Post Office Supplies and W. H. Smith with national distribution networks. Early introduction of the High Speed Train on the Swindon line was important both for business travel back to London and for commuting, whether to Swindon or, increasingly, from Swindon to central Berkshire and London. As Hall et al. (1987) conclude, the M4 corridor and Swindon as a key node within it, 'grew as a result of heavy, if uncoordinated public investment – by a process of disjointed incrementalism, without conscious understanding of its possible consequences'.

In a survey of companies which had relocated to Swindon since the mid-1960s, 64 per cent cited communications as a factor. Growth prospects in terms of markets, land and premises for future expansion were mentioned by 57 per cent, the cost of premises and labour by 22 per cent, and labour availability by 20 per cent (Thamesdown Borough Council, 1983). The M4 and Heathrow were particularly important for multinational companies such as Intel and Raychem. For office-based companies, attempts to restrict office development in London and the south-east using Office Development Permits, further encouraged decentralization to Swindon which lay outside the 'controlled' area.

Plate 2.3   The Windmill Hill Science Park, Swindon.

M4 corridor magnet

With falling birth rates and slower economic growth nationally, the more grandiose 'new city' plans of the 1960s were not fully realized, but there was, nevertheless, major growth in population and employment. The collapse in manufacturing employment and the sharp rise in unemployment in the early 1970s, however, led to policy reassessment locally. In 1977 the council launched a new industrial promotion strategy and set up Swindon Enterprise as a semi-autonomous marketing agency headed by a former commercial director from Plessey. His image of Swindon was a town 'fifty minutes by high speed train from London, one hour from Heathrow, next to the M4 in the golden corridor but cheaper than rival towns, surrounded by stunning countryside with lots of old rectories for executives to live in'.[15] The emphasis was on attracting high technology companies and office headquarters, profitable concerns which did not need development area subsidies to survive. Campus–style business park development was explicitly encouraged through the planning process.

By the 1980s, Swindon occupied a particular position within the south–east labour market based on the juxtaposition of an expanding

[15] Quoted in the *Financial Times*, 30 June 1986.

concentration of basic white-collar, clerical and manufacturing employment, with rural Wiltshire and towns (Marlborough, Bath, Bristol) attractive to technical, professional and managerial groups. Within the M4 corridor, higher-level workers can maximize their employment options. As an insurance products manager, living just outside Swindon and working at the time in Gloucester, described it:

'If I left Imperial Trident, I would certainly think in terms of Bristol, it's close by, or back to Swindon, although there's only Allied Dunbar and NEM. Within striking distance are Cheltenham, obviously Eagle Star, Basingstoke a couple of companies, Andover conceivably which is TSB, conceivably London as well . . . ' (d)

Housing and lifestyle factors are particularly important for such workers. The consequence is that employers located in Swindon can in turn recruit and retain types of key staff who favour a general M4 corridor location.

Expansion from the late 1960s had been rooted in the strategic planning framework for the south-east. By the late 1970s, however, any form of strategic planning at national government level had disappeared. Continued expansion reflected more generalized pressure for growth in the M4 corridor. Recession deepened after 1979. The south-east, however, consolidated its economic dominance within the UK economy, reflected in the growing 'north–south divide'. Expansion of financial and business services reflected London's role in the international economy and there was a growing concentration of newer manufacturing activity and high technology in the region. Partial recovery in the mid-1980s only served to re-emphasize regional imbalance, again working to Swindon's advantage.

New directions?

By the mid-1980s, however, a combination of factors led to major policy reappraisal. This was summarized in the 1984 consultative document, New Vision for Thamesdown (Thamesdown Borough Council, 1984b). The council's long-held claim that continued expansion, new housing and jobs benefited existing residents, rather than simply fostering in-migration, was increasingly questioned within the ruling Labour group and there were some signs of hostility to continued expansion among existing residents. According to the consultation document, the clear vision which had sustained the strategy of town expansion in the postwar years had become increasingly blurred.

In particular, it was felt that the council had lost much of the control and influence which it had previously exercised over the pace and form of town expansion. Central government attempts to control local authority expenditure placed severe financial constraints on the council, and from 1985/86 the council was 'ratecapped' along with overtly left-wing London boroughs and authorities like Liverpool.[16] In Swindon's case, this reflected debt charges resulting from high levels of capital expenditure on town expansion in the past which, ironically, had been explicitly encouraged by central government.

On top of this, the council's landholdings were nearing exhaustion with little prospect of significant acquisitions given not only the new financial constraints but also rising land values which priced it out of the market. The council could no longer, therefore, exercise the degree of control over the development process which it had in the past as a major landowner. Public expenditure constraints posed further problems in terms of infrastructure and service provision. Future development was now therefore largely reliant on the private sector. Against this background, New Vision for Thamesdown recommended continued growth of employment and housing up to the 1990s followed by consolidation or carefully controlled physical growth. It saw the council's role increasingly as one of advocacy rather than direct involvement, depending 'on outside agencies for the powers, resources, employment, investment and much of the infrastructure needed to achieve its vision' (Thamesdown Borough Council, 1984b, p. 61).

Reflecting thinking in the Labour group, a new Directorate of Economic and Social Development was established, reducing the autonomy of Swindon Enterprise. A consultative document in 1987 proposed that the main thrust of the marketing strategy be maintained but with more emphasis on 'new left' initiatives including training, cooperatives, community enterprise, equal opportunities, job quality, and joint planning with trade unions. In practice, high profile promotion of the locality has continued alongside attempts to give greater emphasis to the nature of economic development and to distributional issues.

In terms of physical development, the emphasis switched in the mid-1980s to the town's northern sector, seen for some time as the next major area for development. Debate over the scale and form of

---

[16] The legislation allowed central government to set a ceiling on the rate (local tax) level where a council exceeded specified spending limits. This gave central government tight control over local spending, although 'creative' accounting allowed Swindon and others to avoid immediate spending cuts.

development here illustrates more general issues.[17] In 1986 a consortium of builders submitted a planning application for around 9000 dwellings, representing a population increase of 24,000. Revisions to the Structure Plan proposed by the county but yet to be approved by central government allowed for only 3500 houses up to 1996, with vaguer proposals for continued development beyond that. The consortium, however, was seeking firm commitment to development of the whole area up to 2001, well beyond the statutory planning period. Consultants produced a comprehensive 'development brief' and district plan for the area, usurping in a sense the local authority role and reminiscent of North American 'private sector planning'.

Lacking significant landholdings, local council influence was limited to normal planning powers. Being a major departure from the Structure Plan, the application was 'called in' for decision by the Secretary of State for the Environment and a local public enquiry was held in late 1987. District and county councils were concerned over the lack of public resources for infrastructure and services, although both eventually dropped their opposition in exchange for increased contributions to infrastructure from the developers. Development was opposed by the Northern Action Group composed largely of local residents. The 1986 survey, however, suggested broad-based support for further expansion, at least on employment grounds, with almost eighty per cent of respondents in Swindon agreeing that further expansion would help create job opportunities for local people.

Pressures for housing development in the south-east were intense in the late 1980s, reflecting the region's economic bouyancy. Opposition to further residential land release from established residents had intensified, however, both in Wiltshire and adjoining western corridor counties and in the region as a whole. There was increasing concern over escalating house prices in the south-east and evidence of labour and skill shortage as a result. Given the regional context there was growing evidence that Swindon was becoming in effect a safety valve for the south-east, taking development pressure off other localities. The motor for growth in the late 1980s was thus increasingly residential development as well as economic expansion per se. Swindon was increasingly seen as a lower-priced dormitory area, particularly for Berkshire and London, indicated by the sharp increase in rail commuting.

Development at Swindon was likely to be politically acceptable to a government keen to demonstrate its faith in comprehensive planning

---

[17] Discussed more fully in Harloe and Boddy (1988).

and development led by the private sector. It was, however, caught between pressure from the development industry for land release and vociferous local resistance to further development from predominantly Conservative areas. The problem for the local authorities was one of fulfilling a regional function, but with very limited local resources and powers. Pressure on the Secretary of State to back the development was however strong and approval was granted in 1988. As with earlier phases of development – postwar town expansion and the 'new city' period – the possibilities for Swindon's continued growth again reflected the broader regional context. The nature of these pressures, in particular demands for residential development, and the capacity of the local council to control and influence the pattern of development, were, however, very different by the late 1980s.

Class, politics and cultural change

Economic growth and diversification, and the major influx of population and new employers, have had profound implications for class and social structure, politics and everyday life in Swindon. Social and political change has in turn shaped these processes of economic change. This final section focuses on class and social structure, unions and workplace organization, formal politics and local culture.

Class and social structure

Up to the 1950s, Swindon's social structure, institutions and community life reflected the dominance of male manufacturing employment, in particular skilled engineering employment in the railworks. Rapid employment growth in the 1950s further emphasized this bias towards skilled and semi-skilled engineering, which left foremen and skilled workers significantly over-represented compared with the national picture, and employers, managers and professional workers under-represented. Between 1961 and 1981 however, employers, managers and professionals increased their share of total employment locally from 9 per cent to 17 per cent while the share of foremen and skilled workers fell from 43 per cent to 31 per cent.[18] This brought the town's occupational structure much closer to the national picture. There was thus a marked decline in traditionally dominant strata of male manual workers. But while skilled manual groups, the 'working class aristocracy', declined, semi-skilled and unskilled manual

18 Population Census, Swindon Borough, economically active males by grouped socio-economic group.

workers maintained their share of the workforce, as a consequence of a 'deskilling' of the overall manual workforce.

Under-representation of employers, managers and professional groups had largely disappeared by they 1980s. The 'new service class'[19] expanded with the growth of higher-level administrative and financial functions, new manufacturing and high technology industry. Indicative of the increasing importance of white collar employment more generally, intermediate non-manual occupations increased their share of the workforce from 13 per cent to 18 per cent in the decade to 1981.[20] The number of people with degrees or higher professional qualifications doubled.[21]

Growth of routinized, non-manual labour, much of it in the service sector, has greatly expanded the 'white-collar proletariat'.[22] In a sense, Swindon is still a working class town. The nature of that working class has been transformed, however. This is reinforced by the fact that many managers and professionals working in Swindon live in the surrounding rural villages – small towns like Marlborough or Cirencester and further afield, for example in Bath, Bristol or Oxford. Over a third of professionals and managers working in the district in 1981 lived outside the local authority boundary, compared with 8 per cent of all manual workers.[23] This reflects the residential preferences of managerial and professional groups, and the particular structure of the housing market within Swindon itself. Given Swindon's history there is very little older housing at the middle to upper end of the market and most private housing is on more recent, speculatively built estates.

Unions and the labour movement

Elements of a socialist movement emerged in Swindon from the late nineteenth century based, as elsewhere, on Marxism and Methodism. The Labour Party, established locally in 1916, was strongly supported between the wars and firmly rooted in the unions and the manual working class. There was also support for the Communist Party and the Independent Labour Party. The General Strike was

[19] 'Those occupations, often characterised by a high level of material reward and considerable autonomy within the work itself, that direct, administer and control . . . the major institutions of advanced capitalist society, both public and private'. (Crompton and Jones, 1984, p. 4.) See also Abercrombie and Urry (1983).

[20] Population Census, Swindon TTWA, male and female, social class.

[21] Population Census, Swindon TTWA.

[22] See the extended analysis of Crompton and Jones (1984) which emphasizes the fragmentation of the 'office proletariat' by age and qualification but in, particular, gender.

[23] Population Census 1981, Swindon Borough, defined in terms of social class.

solidly backed in 1926, unionized employment expanded rapidly in the 1950s, and Swindon car workers shared in the rising militancy in the industry from the late 1950s. Economic and social change have, however, largely erased the town's legacy of trade unionism and working class socialist organization, and few Swindon workers would now identify with a tradition of working class politics and industrial militancy.

Craft unions gained increasing strength within the GWR works around the turn of the century. Low wages, rising living costs and unemployment brought mass meetings and petitions in 1908; there were more mass meetings over the introduction of timeclocks in 1913; and subsequent government control over wartime production advanced company recognition of unions. The craft societies grouped under the Swindon Railway Federation declared their support for the abortive Triple Alliance of railwaymen and transport workers in support of the 1921 national miners' strike. The interests of rail engineering workers in Swindon were closely tied up with those of other rail workers and, given company dependence on coal traffic, with the miners. The National Union of Railwaymen was itself strongly represented in the works, which was inevitably drawn into stoppages on the rail system or the coalfields.

In May 1926 support for the General Strike was virtually total and the works closed throughout the strike. The municipally owned tram system was halted and workers in the electricity company, also municipally owned, removed the fuses from the main industrial enterprises such as Garrards (Tuckett, 1976). Strong feelings of solidarity were fostered by the collective experience of the strike. Mass meetings of several thousand were held daily in the GWR park, a mock funeral paraded through the town visiting the homes of non-strikers, and trains passing through the town run by 'volunteers' were stoned. Local support held firm until the strike was called off nationally and the labour movement fought hard to secure satisfactory terms for the return to work, and to prevent victimization. The GWR in particular sought to discipline workers and exclude union leaders and activists in the strike, many of whom were forced to leave the town (Tuckett, 1976).

The strength of the labour movement locally meant that it could exploit the potential growth of unionism which came with the rapid expansion of engineering employment during the Second World War. Under the influence of Joint Production Committees instituted by the government to secure wartime production, full union recognition was won from major employers such as Plessey and Vickers. By the end of the war, as the local paper observed, virtually every worker in Swindon was a trade union member,[24] and employers moving into Swindon in the 1950s were well aware of the strength of the unions in the town. The established strength and organizational structure of the union movement were such that postwar engineering growth brought continued expansion of union membership. This included significant female employment, particularly in electrical engineering which was unionized during the war and increasingly took on women in the 1950s and early 1960s.

The early 1960s was, however, the high point for the unions and the next two decades saw the eclipse of militant labourism in the locality. The car industry, in particular, rapidly established a more confrontational style of management – labour relations. A series of major disputes in the early 1960s saw the employers on the offensive and resulted in significant defeats for local union organization. Swindon car workers went on strike in 1958 and again in 1961, claiming pay parity with Cowley. In 1961 however, formal negotiation procedures had not been completed and union officials opposed the action, exposing those on strike. After three weeks Pressed Steel dismissed the strikers and advertised locally for skilled men. Strikers taken back suffered loss of seniority and pay, known union militants were excluded and new workers were recruited from the north-east. This was described by a former union official as a turning point.

Heavy job loss in the engineering sector in the 1970s then hit unionized employment particularly hard. In 1973, 1200 workers in Garrard, mainly women, were involved in a five week strike against company attempts to intensify production. The company gave in but in the longer term abandoned the industry in the face of Japanese competition. Unions are still recognized by the major employers in engineering and the production side of electronics, and the strength of newer, high tech industry has sustained the craft unions to some extent. They remain, however, more narrowly instrumental and company specific, concerned in particular with wage bargaining and negotiating procedures. At the national level this has been reinforced by anti-union legislation and the 'discipline' encouraged by high levels of unemployment.

Major employment growth in the 1970s and since has been concentrated in sectors characterized by non-union employers. In financial and other office-based services and much of high tech industry companies have successfully sought to pre-empt the need for unions and built a non-union work culture for the mass of their employees. Individual membership is allowed, but unions are not

---

[24] *Swindon Evening Advertiser*, 19 November 1945.

generally recognized for negotiating purposes. In miscellaneous services, retailing, and some elements of clerical and secretarial employment the expansion of part-time and temporary employment has for different reasons militated against unionization. The main exception has been the growth of unionized public sector employment, although this had largely come to an end by the late 1970s.[25]

Local politics

Up to the Second World War, formal electoral politics locally were heavily dominated by the pervasive influence of the GWR. Sir Daniel Gooch, GWR's first locomotive supervisor at the works and later chairman of the company, was the town's MP from 1865 to 1885. Swindon's first mayor was locomotive superintendent at the works and by the early 1920s up to nineteen of the thirty-six local councillors worked for the GWR, including manual and clerical workers and management. The town's petty bourgeoisie – the small traders, professionals and shopkeepers – represented on the council were in any case largely dependent on the fortunes of the town's dominant employer.

Between the wars 'Citizens' League' members and, later, 'Independents' held the majority on the council. Though politically conservative, they were not affiliated to the Conservative Party, did not meet as a formal group, and tended to be strongly 'localist' in outlook. The Labour Party had four seats by 1920 and made significant gains between the wars. It was not until 1945, however, that it won control of the council, major gains reflecting the strength of the unions and labour movement locally – and, indeed, in much of the country at this time. Independents regained power from 1949 to 1951 and remained the major opposition group right through to 1968.

After the war, the locally oriented Independents contributed to strong non-partisan support on the council for town expansion. The policy itself was developed and implemented by a relatively small group of Labour members and key officers, in particular the town clerk, David Murray John. They were, however, able to sustain wide support both within the council and from a broader range of interests locally, including the unions and rail works management. This pro-growth coalition represented a form of territorial alliance with some similarities to the spatial coalition on Teesside in the 1960s

---

25 The 1986 SCEL Employer Survey indicated that employees were not represented by unions in 51 per cent of establishments (58 per cent in the case of larger private sector establishments, 79 per cent for smaller private sector establishments (less than twenty employees) but only 4 per cent for larger public establishments). Only 15 per cent of larger private sector establishments reported that unions were 'encouraged' compared with 45 per cent of larger public establishments.

committed to 'modernization' (see Chapter 8), and support for economic development policies in Lancaster (see Chapter 4). The coalition reflected political patterns established during GWR dominance and the continuing railway presence on the council. With the threat of declining employment in the railworks, the unions argued strongly for the retention of wartime industry and diversification of the town's industrial base, while management saw that expansion would facilitate anticipated cuts in the workforce. Expanding companies, in particular Vickers and Plessey, experiencing labour shortages, were keen to see the town's workforce expand, and urged the provision of local authority housing, though they did not favour increased competition from new employers.

Later years saw local politics increasingly organized and fought on national party lines. The Independents were rapidly eclipsed by the Conservatives in 1968, the latter emerging as a largely new political grouping rather than one made up of former Independents adopting a Party label. This happened much later in Swindon than in many other localities where there was a major shift from Independents to Conservatives in the early 1950s. Interesting contrasts can be drawn with Ashford, Kent, also an 'expanded' town and also rooted in rail engineering (Harloe, 1975). There, the local Independents were eclipsed by Conservatives in the early 1950s and the Conservatives were able to politicize and mobilize middle class opposition to expansion over a crucial period. In Swindon, by the time the Conservatives gained control, town expansion was an established success.

Many of the Conservatives were younger and represented the newer industries moving into the town, for whom ideologically motivated opposition to the Wilson government was more important than local issues. Conservatives gained control of the Council from 1968 to 1970. Labour, however, regained control in 1971 and have retained it since then, except for a brief period when they lost overall control. The 1970s saw the collapse of blue-collar, unionized, manufacturing employment, rapid expansion of white-collar, service occupations and strong growth of female employment, particularly part-time. Labour Party support was not, however, undermined, as might have been expected. Support for Labour, at district council level at least, increased strongly from the late 1970s and the party established its largest majority ever (of twenty two) in 1986.

In part, this reflected the expansion of public sector employment, the 'white-collar proletariat' and manual labour in warehousing and distribution – groups generating continued support for Labour. Other more specific factors also played a part. Labour councils have been identified with successful town expansion, job creation and the provision of physical and social amenities: 'whatever you might say about a Labour Council, Thamesdown Corporation has done a remarkable job over the past ten years or so in promoting Thamesdown' (d). Similarly, 'a lot of people see the good that the Labour authority is doing, things like the sporting facilities, like pressuring Wiltshire to provide schools' (h). There is some evidence that potential Alliance or Conservative voters have identified their personal interests with a local council committed to public service provision. Finally, many potential Alliance and Conservative supporters working in the town actually live beyond its boundaries, so that politically the district council area is more 'working class' than its overall employment structure would suggest.

The Conservatives did, however, win the Swindon parliamentary seat in 1979 – a surprise result in a long-held Labour constituency. The Conservatives retained it in 1983, and again in 1987 with an increased majority, suggesting a more basic shift in the constituency's political base. The contrast with Labour strength at district council level partly reflects different boundaries. The Alliance gained three seats on the district council in 1987, two from Labour, though whether this is indicative of any longer-term trend is not yet clear. The view is commonly expressed locally that social change must eventually tell against Labour: 'I think they've made a rod for their own backs' (e) and:

'Labour are in grave danger of doing themselves out . . . the prices these houses are going for . . . you're getting commuters to London and they ain't going to vote Labour. One of these days they'll get ousted. As it expands it will become a Conservative town.' (f)

Similar views have however been expressed throughout the postwar period without being realized.

By the mid-1980s a shift was evident within the Labour Group, with the election of younger, more ideologically motivated councillors. The old 'municipal entrepreneurialism' which had carried Labour councils through the 1960s and 1970s began to be questioned by those with perspective closer to the 'new urban left'. The divisions were however, much less sharply drawn than in many Labour councils which shifted to the left in the 1970s. There were signs of change in the Conservative Party as well, with a more Thatcherite outlook increasingly evident.

Thus there was some sharpening of the party political divide in the 1980s, although ambiguities remained. Conservative councillors and the MP Simon Coombs gave at least token support to the campaign against the closure of the rail works, though Labour and the unions made the running on this. In 1985–86 local Conservatives called on Labour to sell land and freeze recruitment to avoid ratecapping, and investigated the possibility of a deal with the minister in return for sale of council assets, a move angrily condemned by Labour. Even so, Conservative views were not clear-cut. In Parliament, Simon Coombs supported ratecapping but argued that Swindon was a special case; the chamber of commerce attacked the government decision; and when Labour voted to reaffirm its budget in September 1985, putting it in line for ratecapping, five Conservatives abstained rather than vote against the budget. Nor was the Labour group entirely as one. Councillors agreed at a conference of ratecapped authorities to collective defiance by refusal to levy a rate. Other councils caved in, however, and in the midst of acrimonious internal debates, Swindon Labour group agreed by one vote to set a rate. Town expansion itself became more of a political issue in the mid-1980s; there was controversy, too, over the change in emphasis in economic development policy and the appointment of the new director, formerly economic adviser to Merseyside Enterprise Board and a past Labour parliamentary candidate. The Conservative leadership thought they were seeing the 'thin veil of moderate socialism being torn aside'.[26]

Other social and political movements remain relatively poorly developed locally, in contrast to the more pervasive and robust working class political culture of the interwar period. There have been networks of activists and specific campaigns but these have tended to rely on relatively small numbers of people. The absence of a university or polytechnic seems to have been an important factor when Swindon is compared with Lancaster, for example. The women's movement in Swindon has developed specific initiatives and campaigns but has drawn on a relatively narrow base and much smaller numbers than have been mobilized in Lancaster. The formal political arena remains largely male dominated. There have been organizational changes within the council and increased commitment to issues of gender, race and equal opportunities, including the appointment of an officer in the Community Development Department with responsibility for women's issues. Such moves largely originate, however, within the council, reflecting the rise of the new left and new thinking at officer level. The announced closure of the railworks led to the 'Railworks Defence Campaign', supported financially and organizationally by the Labour council which actively backed the unions. Oriented essentially to fighting closure, the campaign failed to mobilize mass support, had little real power to oppose British Rail's decision, and was thus inevitably somewhat tokenistic. Some argued, with the benefit of hindsight, that it might have been better to direct the campaign towards securing the best deal from British Rail in terms of land and facilities for local use.

[26] Quoted in *Swindon Evening Advertiser*.

Culture and locality

Life in Swindon up to the last war revolved around the railworks: 'just everybody used to work in the railway. GWR was Swindon and that was it' (s). The heavily structured work culture and the rhythms of activity in the works permeated the community to which it gave birth (Williams, 1915; Peck, 1983). Both daily life and major social occasions centred on the works. The annual children's fete, which continued up to the last war, was attended by 34,000 in 1904. In 1905, the annual holiday week, 'the trip', saw 24,500 people, almost half the population, leave the town on 22 special trains. This dominance declined only slowly in postwar years and the feelings of older residents over the final closure of the works in 1986 reflected the loss of something much more than simply jobs:

'From a personal point of view, I think we've lost our heritage, because I was born when Swindon *was* railway and I believe that that was real, and I think the high tech is plastic. I think it's made possibly a generation of 50-year-olds feel a bit useless really . . . they were skilled workers . . . very skilled . . . I think the suit and briefcase lady is coming to be. . . . We've lost our culture, the railway culture was a way of life.' (g)

Popular identification of the town with rail engineering remained strong even in the 1980s. Right through from the nineteenth century to the 1980s, if you worked in the railworks you were 'inside'. Over half the respondents to the 1986 survey still associated Swindon closely with railways and rail engineering. The dominant feelings, however, were regret and nostalgia – 'it's sad because we're talking about tradition, heritage' (m) – commonly coupled with more hard-nosed realism: 'Everyone says it's a shame about the railway, and it is. But at the end of the day it's something that had to happen. If it's not making money, it had to go'. (o).

Postwar expansion greatly increased the number and diversity of employers. There was much less overlap between peoples' work and non-work lives, and the growing population had much less in common in terms of work experiences. With massive in-migration from London and elsewhere, much of the town was marked by the newness of the population and the lack, initially at least, of extended social networks. There is still awareness of the contrasts between Swindonians and newcomers, particularly Londoners, arriving in the late 1950s and early 1960s:

'When the Londoners first came and they didn't like this and they didn't like that, then I used to get so upperty about that. I used to think if you don't like it, go back to where you came from. Now I know quite a few Londoners. They're quite nice once you get to know them.' (r)

And the feelings were often mutual: 'We were very much resented when we arrived, "damned Londoners taking all our jobs" for a very long time' (l).

> 'There was a lot of resentment when we first moved down here because regardless of who you were, what you were, you were all classed as Londoners and classed as outsiders. That could be difficult when you first came. I mean, it was like a village, you couldn't call it a town.' (c)

Only about a third of the 1986 population was born locally and a proportion of these would be children of migrants. Stereotypes of Swindonians and Londoners remain strong, however. Swindonians will be described as 'reserved', 'slow to accept you', but 'fine once you get to know them'. Differences are also perceived between 'Londoners' and 'the new people', the more recent newcomers of the 1970s, 'a different kettle of fish ... it's two completely different towns now' (t).

> 'Swindonian people were really working class people, and I think that the people that have come from outside tend to be, as I said, a bit up market.... The new people are different from the Londoners who came in the 1950s who were also working class which helped them integrate.' (s)

The influx of migrants is associated with a (romanticized) loss of community: 'the new people coming don't seem to think of it as a community. They think of theirselves, people nowadays, as long as I'm all right, that's all that matters'. (s) The contrast with the collectivity of the interwar period focused around the rail works is stark: 'once everybody knew everybody, now nobody knows nobody'. (s)

The increased privatism and home-centredness typical of many places in later postwar years was emphasized in Swindon's case by the newness, the isolation of young families, differences in background and the dislocation of social and family ties consequent on in-migration:

> 'Initially, women were likely to be preoccupied with child-rearing ... by and large they are young married people with families. The men go to the factories, have a beer, watch telly; the women bring up the children. On the estates, they are still in the introspective stage'.[27]

Continued in-migration through the 1970s and since has reinforced this predominance of individualized lifestyles: 'it's not unfriendly, but everyone keeps themselves very much to themselves' (p). In the Western Expansion

[27] J. Rogaly in the *Financial Times*, 27 May 1966.

'people are very wary about knocking on your door and saying, "Hello, I'm your new neighbour, do you want a cup of tea".... There are a lot of people who live in this street who leave at seven to eight in the morning, travel to work, work, get home seven to eight at night, so you're knackered.' (h)

There is considerable awareness, as well, of pressures of debt, consumerism and the need to keep up appearances: 'This is a successful area, so you have got to be successful.' (h) 'The majority are mortgaged up to the absolute hilt. From the outside you can look at people and think that they're probably doing extremely well, but you don't realize they are probably up to their eyes in debt . . . they are so materialistic.' (h) With the increase in the number of women working, women are now much less likely to be preoccupied by the home and child rearing – although, according to the 1986 survey, this appears to have had little impact on traditional gender divisions of domestic work: over 80 per cent of women still cook the main household meal and nearly 90 per cent do the ironing.

Shopping and leisure provision are thought of as very good, though 'It's what I would class as a utility town. It's just got the basics here, your normal Marks and Spencers, BHS, C&A, it's not big enough ... you would have to go to Bristol or Bath for anything special.' (f) Cheltenham and Oxford are alternative shopping venues for 'anything a bit different'. Young people and younger single professionals often find Swindon boring, with little to offer by way of entertainment or excitement. Sports and recreational facilities are widely seen as excellent, 'second to none' even: 'We've got an abundance of leisure facilities in Swindon, there are community centres and sporting paradises being thrown up all over.' (q) But 'There's not really much left in Swindon if you're not into sport'. (j) For those with more resources, 'The town's OK for your standard Berni and steak houses, but that's not always what you want'. (q) Apart from the many pubs, legacy of an earlier male dominated working class culture, evening entertainment is limited and again, Bath, Bristol or Oxford are alternative venues for some. The town's housing, environment and facilities are widely praised: 'the ideal place for the average family with a couple of kids', though for managers, professionals and higher grade employees it is described as 'soulless and lacking in social cachet'.

Swindon is not in any sense a 'cultural centre'. It lacks any historic, religious or educational roots beyond the latter half of the nineteenth century. The 'Brunel' shopping centre echoing railway architecture, renovated railway cottages and the railway museum are reminders of the recent industrial past, but the contrast with places like Bath or

Plate 2.4 The Swindon Link Centre: a community leisure complex.

Bristol or 'towns with character' like Cirencester and Marlborough is very marked. The civic Wyvern Theatre seldom ventures beyond pantomime and worthy repertory, and is contrasted unfavourably with Bath, Bristol or Oxford: 'Well, there's the Wyvern Theatre, but . . . ' As one personnel manager put it, in terms of attracting higher-level staff 'Swindon still has an image problem', although less so than in the past. Some companies have advertised themselves as located in the M4 corridor or Wiltshire rather than Swindon itself. One manager said, in relation to recruiting people to Swindon, that his company 'pretend we're not here'. Company and council promotion stresses the surroundings, the rural villages, historic centres and countryside rather than the town itself. According to a resident of Highworth, three miles beyond the edge of town, who goes to Swindon 'as little as possible':

> 'I suppose one must bring class into this, which I didn't want to do. Highworth is a relatively well off area. From a marketing point of view most people are As, Bs, possibly C1s round here. . . . ' (d)

And a resident of Purton, three miles west of the town, commented: 'I don't have much to do with Swindon at all'. (f)
  In terms of urban fabric, there are references back, reconstructions

and pastiche: the renovated railway village cottages, the railway museum and the 'Brunel' shopping centre have been mentioned. The Western Expansion and planned northern sector development have self-consciously adopted an 'urban village' format. And the planned Tarmac development on the railworks site incorporates high density, disorganized 'village type' housing developments along with 'museumification' of retained buildings.

This postmodernist emphasis on heritage and reconstruction is overshadowed, thus far, by late modernist high tech, the main contender for the dominant image of the locality. Over three-quarters of respondents to the 1986 survey agreed that the town is closely associated with high technology and computers, despite the small percentage of employment actually in high technology defined on the basis of official statistics. The high tech, M4 corridor image has been consciously cultivated by council policy and, importantly, by the private property sector seeking to sell new developments. The council has used the pressure for development and relocation to secure high design standards for both residential and commercial development. Archetypal 'high tech' architecture strongly projects this image of modernity. It includes specific buildings such as the silver and glass Murray John Tower dominating the town centre and symbolically named after the former town clerk, the nationally famous Foster-designed Renault building, futuristic campus-style developments, and the town centre office cluster, a 'mini-Manhattan', dominated by Allied Dunbar.

Things can look very different, however, from the perspective of the seven thousand unemployed, a group which fits uneasily with the images of expansion and modernity. As one unemployed 23-year-old saw it:

'it sounds pretty daft, but in Swindon at the moment, they seem to be recruiting a lot of people from outside, especially with skills, and bringing them all in, and unemployed who actually have lived here virtually all their lives, they are finding it difficult to get work because it's all specialized industry' (k).

Over 75 per cent of people interviewed in the 1986 survey thought that it was no different being unemployed in Swindon than elsewhere. Of those who thought it was different, the majority suggested it was easier to find work, and the feeling that people can find work if they really look for it seems to be stronger in Swindon than in many places. Attitudes to unemployment seem different compared with areas more accustomed to high levels of unemployment: 'there are so many unemployed up at home, in Sunderland, that it is almost the norm, it's accepted, that, and I think that in Swindon, it's an unusual thing not to be working'. (h) The poverty and alienation are if anything reinforced:

> 'it's claustrophobic, everybody comes into town, and there is not that much to do within the town itself. So people find it quite boring really . . . It gets fairly depressing because there is sod all to do in the evenings, unless you have money to go out to the pub . . . there isn't a community feel. A lot of people in the street you will smile at, and don't get any response back, and others have said this too. . . . I still think it gets down to the growth of the place and how many new people are coming in.' (k)

There is little sense here of the solidarity and support within the community developed in places like Kirkby, as described in Chapter 6. The town's overall prosperity meant little to many of the unemployed, for whom poverty was the dominant experience. Nearly two thirds of a sample of two hundred unemployed in the 1986 survey found it hard to make ends meet. Thirty-four per cent found it 'very hard' compared with only 3 per cent of those working full time. Moreover, over a quarter thought it very unlikely that they would find a job in the next year.

Conclusions

Swindon's economic and social transformation demonstrates the central importance of the relationship between change at locality level and the restructuring of the broader urban and regional system, itself inextricably bound up with the shifting nexus of forces at national and international levels. Through successive stages of development, the location of the town, particularly in relation to the south-east region, has been crucial. The nature of this relationship changed, however, through successive periods of economic and social development: the town's strategic location on the GWR network, its location beyond supposed bombing range in the Second World war, postwar town expansion based on overspill from inner London, the 'new city era' linked to broader regional growth pressures and strategic planning policy and, most recently, pressures for economic and social development which reflect the town's nodal position as a key growth centre within the M4 corridor and its integration into the increasingly dispersed urban and regional system of the south-east heartland.

Successive waves of development built on and reworked what had gone before. Wartime relocation of manufacturing and the postwar boom in engineering built on the tradition of skilled engineering. Later service growth drew on the availability of female labour. The different stages of growth and development were, however, to some extent independent in origin and overlaid one on the other. It is more than a truism that, but for its location, Swindon might be little different now from the northern industrial towns which it resembled up to 1945, but which were hit by the collapse of traditional manufacturing sectors. It is the succession of rail engineering, wartime industrial production, postwar engineering boom, growth of services and distribution, and, lastly, M4 corridor growth that is distinctive to Swindon.

At one level, then, the locality is increasingly integrated into the regional and indeed national and international economy and labour market. Internally, however, the picture at locality level is one of increasing fragmentation and complexity, with new employers, new forms of economic activity and new forms of work. And the locality bears little resemblance to the revitalized, spatially integrated 'new industrial districts' based on subcontracting relations and the need to respond to rapidly changing markets, put forward by some writers (Sabel, 1987; Scott, 1988).[28]

The particularly active role played by the local authority in the case of Swindon helps clarify the broader issue of 'pro-active capacity'. Local policy and the particular configuration of local interests in Swindon were necessary factors determining the pace and form of town development. Local policy sharply differentiated the locality from neighbouring areas and indeed much of the outer south-east region. Pro-active capacity should not, however, be thought of as some form of independent ability of local policy or interests to achieve their goals. It reflects, rather, a specific combination of local factors and wider context. The 'pro-active capacity'' of the local authority can only be understood in terms of Swindon's place in relation to the south-east regional economy. It lay in the council's commitment and ability to capitalize on the town's growth potential, afforded by its position within the changing urban and regional system of the UK and the consolidation, within this, of the south-east region as the country's economic heartland. Policy at the level of the nation state has also played a key role at different points in time, including planned wartime relocation of industry, reconstruction, protectionism and its demise, as well as specific regional policy and strategic planning. Essentially uncoordinated policy decisions on

---

[28] See Lovering (1987) for critical discussion of these arguments. Information from the 1986 SCEL Employer Survey confirms the non-local emphasis in linkages: only 30 per cent of larger private sector establishments had local customers whereas 50 per cent had national and 53 per cent international customers. Fourteen per cent included local suppliers among their main sources compared with 46 per cent who drew on national and 39 per cent international sources.

infrastructure have nevertheless combined to consolidate growth within the western corridor and Swindon's place within this. One can also point to the role of government in supporting and facilitating London's role as a key international financial centre as contributing to continued growth pressure within the south-east. With increased integration into the south-east economy and increasingly centralized government control, the role specifically of local policy, and the scope for local pro-activity, has, however, been undermined. Pro-active capacity has contracted and, in effect, shifted from the public to the private sector with very different implications for the 'successful' realization of continuing growth potential.

Finally, Swindon demonstrates particularly rapid change in social and cultural terms, which has reflected but also shaped the pattern of economic restructuring and the recomposition of the workforce. Support for town expansion which encouraged economic growth and diversification, for example, was rooted in a particular configuration of social and political interests locally. The social and cultural associations of surrounding rural areas and smaller urban centres has been a key factor in ensuring the availability of professional and managerial labour. And the local authority, again, has played an important role in fostering and providing the social and physical infrastructure for an expanding supply of more basic manual, clerical and white-collar labour in the town itself.

As at the economic level, in social and cultural terms the town has become increasingly fragmented and complex. Working class solidarity, collectivism and community ties have given way to home-centred individualism and social fragmentation. In a sense this reflects the growing homogeniety of local and regional culture and ways of life. The locality has become more like other growth centres in the outer south-east. But this internal fragmentation is all the more marked because of the contrast with patterns of life associated in the past with rail engineering, the scale of expansion and in-migration and the pace of economic transformation. Once everybody, it seemed, was 'inside'. Economic and social change, however, has turned the town inside out.

References

Abercrombie, N. and Urry, J. (1983) *Capital, Labour and the Middle Classes*, London: Allen and Unwin.

Atkinson, J. (1984) 'Manpower strategies for flexible organisations', *Personnel Management*, August.

Crompton, R. and Jones, G. (1984) *White Collar Proletariat: Deskilling, and Gender in Clerical Work*, London: Macmillan.

Hakim, C. (1979) 'Occupational segregation', Department of Employment Research Paper 9.

Hall, P., Breheny, M., McQuaid, R. and Hart, D. (1987) *Western Sunrise: the Genesis and Growth of Britain's Major High Tech Corridor*, Hemel Hempstead: Allen and Unwin.

Harloe, M. (1975) *Swindon: A Town in Transition*, London: Heinemann.

Harloe, M., and Boddy, M. (1988) 'Swindon: A suitable place for expansion', *Planning Practice and Research*, September, pp. 17–20.

Hudson, K. (1967) *An Awkward Size for a Town: A Study of Swindon at the 100,000 Mark*, London: David and Charles.

Lovering, J. (1987) *Britain's Changing Urban Structure and the Effectiveness of Local Economic Policies*, paper presented at symposium on 'Declining Cities and Urban Politics', University of Bremen, October.

Peck, A. S. (1983) *The Great Western at Swindon Works*, Poole: Oxford Publishing Company.

Sabel, C. (1988) *The Re-emergence of Regional Economies*, paper presented to colloquium on 'Logiques d'enterprise et formes de légitimité', Paris, January.

Scott, A. (1988) 'Flexible accumulation and regional development: the rise of new industrial spaces in North America and Western Europe', *International Journal of Urban and Regional Research*, 12, pp. 171–86.

Thamesdown Borough Council (1983) *How Many Jobs? A Study of Direct and Indirect Job Creation*, Swindon Enterprise.

Thamesdown Borough Council (1984a) *Employment Information Report*, Corporate Planning Unit.

Thamesdown Borough Council (1984b) *A New Vision for Thamesdown*, Consultation Document.

Tuckett, A. (1976), 'Swindon', in J. Skelley (ed.), *The General Strike, 1926*, London: Lawrence and Wishart.

Williams, A. (1915), *Life in a Railway Factory*, Gloucester: Alan Sutton.

Appendix 1: Key to interviews

(a) Twenty-eight-year-old female clerical worker.

(b) Twenty-one-year-old female clerical worker, in financial services.

(c) Fifty-eight-year-old female clerical worker, recently made redundant in retail sector following computerization.

(d) Thirty-nine-year-old insurance products manager living in Highworth, a village just outside Swindon, and working in Gloucester.

(e) Thirty-four-year-old woman, who moved to Swindon from London with parents and lived in Swindon most of her life.

(f) Thirty-two-year-old civil engineer living since 1983 in Purton, a village just west of Swindon.

(g) Forty-four-year-old nurse, working part-time, who has lived in Swindon all her life. Her husband works for British Rail as a train driver.

(h) Thirty-four-year-old unemployed man living in Western Expansion whose wife works full-time, and who has a history of paid community work. Originally from Sunderland, he moved to Swindon in 1985.

(i)   Fifty-seven-year-old, single unemployed man who had lived in Swindon all his life.

(j)   Unemployed twenty-two-year-old male, who had lived in Swindon all his life.

(k)   Male in early twenties, came to college in Swindon from North Wales and has been looking for work since finishing his course.

(m) Male field services manager who moved to Shrivenham, outside Swindon about nine years ago.

(n)   Accountant, male, aged fifty-one who moved to Swindon with his company in 1973.

(o)   Twenty-seven-year-old part-time shop-worker who had lived in Swindon for twenty-one years.

(p)   Twenty-five-year-old full-time female clerical worker who moved to Swindon from the Southampton area in 1985.

(q)   Thirty-six-year-old female clerical worker who had lived in Swindon for six years.

(r)   Fifty-two-year-old full-time female clerical worker born in Swindon. Her husband was made redundant from the rail works in 1986.

(s)   Fifty-two-year-old female working part-time as a cleaner in a private nursing home, who had lived all her life in Swindon.

(t)   Male press operator at Austin Rover, where he has worked for over twenty years. He left the rail works in the early 1960s at the time of the 'Beeching' cuts.

3

Cheltenham: Affluence Amid Recession.

HARRY COWEN, IAN LIVINGSTONE, ANDY McNAB, STEVE HARRISON, LAURIE HOWES & BRIAN JERRARD[1]

Introduction

The name of Cheltenham Spa has long conjured up images of colonels, spa water, leisure, festivals, the Ladies' College and Regency architectural grandeur (Little, 1952). These images are not unfounded. Situated at the foot of the Cotswolds, its scenic surroundings have proved to be a constant source of attraction for tourists. But the current Cheltenham area of some 106,000 people (Cheltenham Policy Area, OPCS, 1981) bears only a deceptive resemblance to the picture described above. Such is the suggestive power of human artefacts. In the 1980s the locality has constituted an archetypal green-field economy in one sense, but a specific culture in another, which has depended not only upon a diverse industrial structure, but also on national state connections and on the locality's unique ambience.

The fundamental purpose of this chapter is to demonstrate how such elements have been intertwined with the ideological goals of global restructuring and Thatcherism. We also delve beneath the surface of images which have been transmitted as unambiguous symbols of the locality's public culture. In the process we hope to trace the genesis of such images and their significance for contemporary change in Cheltenham. This involves looking at Cheltenham's longer-term history, at the creation of its industrial structure and at the more recent changes that have transformed its economic and social contours. A key theme running through such an inquiry will be the interplay between the locality's unique ambience and industrial restructuring, with its more global reference points.

[1] School of Environmental Studies, Gloscat, Oxstalls Lane, Gloucester.

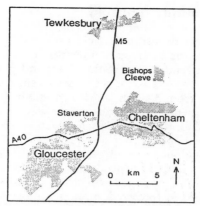

Figure 3.1A Urban centres in north Gloucestershire.

Figure 3.1B Administrative and statistical limits to the Cheltenham locality.

Cheltenham's modern history

Cheltenham's modern history has been marked by periods of rapid growth, including the latter years of the eighteenth century, the early years of the nineteenth century, the 1940s and, more recently, the 1970s and 1980s. Following the discovery of spa springs, the town's population grew from 3000 in 1801 to 31,000 in 1841, an era when the distinctive Regency built environment was created. It developed as a town planned for the *rus in urbe* sense of space and locality, with its terraces, pump rooms, squares and promenade. As the nineteenth century created colonial wealth on an unparalleled scale, affluence spawned its stark opposite, dependency and squalor. In the process, a brief radical culture emerged which emphasized working class dignity. From 1832 to 1865, radical, non-conformist, liberal politics were vigorously maintained, arguably as a response to the strong toryism of the evangelical Reverend Francis Close (Hart, 1981). Close founded the Cheltenham Boys' College which prepared boys for military and civil service in India, thereby reinforcing the linkages between Cheltenham and India, from where army officers, administrators and East India Company merchants had retired for the medicinal waters. He was also largely responsible for founding both the Ladies' College and the St Paul's teacher training college. Thus the imprints of the Church, the military and a social class elite were stamped on the area's educational fabric. During the early years of the twentieth century to 1914, electoral power in borough, county and parliamentary constituencies was shared between Liberals and Conservatives. An Independent Labour Party, set up in 1896, also coexisted alongside the Liberals (Hart, 1981; Hurley, 1979; Pakenham, 1971). Otherwise, political representation in twentieth century Cheltenham has been dominated by Conservative MPs and councillors until the 1980s. The Labour Party contested all the wards for the first time in May 1961, and by 1971 Labour held eleven council seats, the Tories twenty-two, the Liberals four and Independents three (*Gloucestershire Echo*, 1961, 1971).

The rise of industry

Even by 1900, Cheltenham could scarcely be called an industrial town, in spite of brewing and some craft trades (Little, 1967, p. 2). Most earlier Gloucestershire manufacturing activity – the woollen industry, for example – had developed in other parts of the county, such as Stroud and Cirencester. Cheltenham's own economy had been more reliant upon the town as a resort for the rich in the nineteenth century, a dependence which led to a host of employment problems for the local labour force, many of whom were agricultural labourers on the surrounding farms. But domestic service was also a major source of employment for working class families. Such features, symptomatic of Cheltenham's late industrial development, help to explain the paradoxes of contemporary Cheltenham's economic, physical and social fabric (Cowen, 1987a).

It was the First World War that created the embryo of Cheltenham's manufacturing structure in aircraft production, a structure that was to transform the whole employment profile of central Gloucestershire. A former craft iron and cabinet-making firm, H.H. Martyn Aircraft, began operations in a factory west of the town centre, and during the war many Cheltenham workers were employed on aircraft subcontract work (Whitaker, 1985). After 1920, a family member, A.W. Martyn, formed the Gloster Aircraft Company, employing a pool of highly skilled craft engineers that had developed during the war years. The area matured industrially. Firms such as Smiths Industries, manufacturing electric clocks, arrived in the 1920s, locating in the Borough of Cheltenham. In 1926, Gloster Aircraft opened its factory at Brockworth, close to Gloucester, manufacturing a series of well-known aircraft.

At the end of the First World War the Cheltenham Council invested in its first council housing stock. Six hundred council houses at St Mark's were provided for Smiths and Martyn workers as part of a strategy to entice new industry into the area (Pakenham, 1971, p. 142; Hurley, 1979, p. 121). 'Factory Sites! Why not Cheltenham?' appealed one of the posters in 1931 (Beacham and Blake, 1984). The area's environmental attractiveness and the growing pool of trained engineering labour were factors that undeniably drew industry. But the immediate proximity of the town's agricultural hinterland and the fact that in the interwar years many of Cheltenham's population were working in agriculture and service industries ensured a low-wage climate conducive to light industry (Hurley, 1979, p. 117).

George Dowty, a former Gloster Aircraft employee, set up in Cheltenham and by 1939, the Dowty Group employed 3000, expanding through a series of small factory premises on the Severn Vale towards Tewkesbury and Gloucester. In the 1930s Walker-Crossweller came to Cheltenham, as did Rotol (eventually to combine with Dowty) and other, smaller precision engineering firms. Labour was drawn to Cheltenham from depressed areas such as south Wales (household interview). Alongside the industrial expansion, housebuilding activity increased significantly, especially local authority housing for key company workers. By the beginning of the Second World War more than 2000 council houses had been built, in addition to private suburban and infill residential developments. In each case, the estates were developed adjacent to the factories (Hurley, 1979, p. 128).

Thus, with much of the United Kingdom experiencing the world's worst ever economic depression, the Cheltenham economy grew, much of its industrial structure relating to aircraft engineering. The war years quickened this trend. Again, for strategic wartime purposes, the multinational Smiths Industries Aerospace and Defence operations were steered from London to a green-field site close to nearby Bishops Cleeve village, which was soon extended by the provision of Smiths Industries company houses. After the war, overemphasis on defence engineering produced employment difficulties in a context of military run-down. Dowty began a major restructuring of its activities away from aircraft production; Gloster aircraft was rationalized in Gloucester in the 1950s.

But the industrial events of the decade seem highly instructive for understanding the respective fortunes of Gloucester and Cheltenham to the present day. While the Gloucester vicinity received mass product, mass market industries and chemicals, the Cheltenham area received government services and producer services in addition to precision engineering companies such as Spirax Sarco and Telehoist. A crucial state relocation was the move of the Government Communications Headquarters and the Joint Technical Language Service to former Foreign Office buildings (erected after the Second World War) in Cheltenham in 1951. This brought thousands of jobs to the locality. As for other major employers, special local authority housing was constructed – this time five hundred houses and five hundred flats financed by central government in the 1950s on an estate adjacent to what is now the Benhall GCHQ site (Hurley, 1979, p. 177).

The trend towards service industries in Cheltenham continued in the 1960s, with the government's strategic policy of office decentralization from London, beginning with the Eagle Star move to Cheltenham in 1966. A number of other companies followed in the 1970s, virtually forging a completely new sector of financial services and insurance companies in the area (McNab, 1987). Other forms of labour were imported into the town in their wake.

Industrial structure and restructuring in the 1970s and 1980s

Unlike the majority of other UK areas, Cheltenham's economic fortunes once more flourished during the 1971–81 decade which witnessed a restructuring of the UK capitalist economy – total

Table 3.1 *Percentage change in persons economically active, Cheltenham TTWA, 1971–1981*

	1971	1981	% Change 1971–1981
Male	40380	42251	+ 4.6%
Female	24327	28292	+ 16.3%
Total	64707	70543	+ 9.0%

Source: OPCS, 1971, 1981

employment in the Cheltenham travel-to-work area grew by one-third as against a decline of 2.5 per cent in Great Britain as a whole. Such a disparity has to be explained in the light of national and international economic restructuring processes and the changing spatial division of labour in the UK (Massey, 1984) within which Cheltenham is playing a distinctive role.

The globalization of production and markets has intensified the restructuring of the locality's engineering. Similarly, the internationalization of financial markets (Thrift, 1986) has led to internal reorganization, and the rapid spatial concentration and expansion of business and producer services in the south of England (Martin, 1986, p. 283).

The system of production in the UK is also being guided by a new flexibility reflected by the emphasis on all-round high technology skills and formal qualifications in Cheltenham's precision engineering firms, management efficiency informed by international practices, and the systematic utilization of subcontracting to minimize the impacts of competition.

State political processes impinging on the Cheltenham economy may be characterized by the neo-conservatism of the monetarist and privatization strategies of the Thatcher government (Miliband et al., 1987) embodying a restructuring of public expenditure to shore up defence and de-emphasize social spending on housing, health and education. The Friedmanite free market philosophy vociferously advocated by Sir Keith Joseph in the 1970s has facilitated the acceleration of privatisation in Cheltenham's further education provision, with college portfolios increasingly wedded to local business; the liquidizing of land assets to fuel private property development markets; and the ebullience of independent, fee-paying schools and colleges. But the free marketeering of the Thatcherist era has been matched by the strengthening of state control (Whitaker, 1987) – a statism channelling resources into surveillance as part of the 'enormous expansion in the administrative "reach" of the "state"' (Giddens, 1984, p. 184), but simultaneously furthering the economic

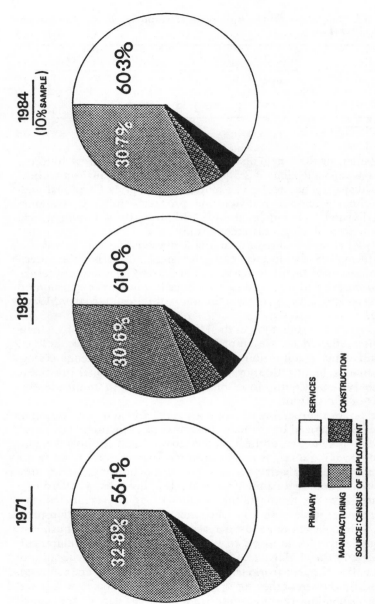

Figure 3.2 Employment in Cheltenham travel-to-work area by industrial sector, 1971–1984.

and social divide between a hub of state such as Cheltenham and a deindustrializing Merseyside economy, for example, from which state military functions were removed after the Second World War.

Finally, the feminization operating across the UK economy has been a multi-faceted process in Cheltenham. On the one hand it has led to a more secure full-time clerical workforce in the producer services sector, a more highly qualified administrative staff in the public sector agencies, and a 'flexible' part-time teaching labour pool. On the other hand it is sustaining the insecure, low-income labour force of the locality's consumer services sector.

By 1984, the distribution of employment by sector in the Cheltenham area was as shown in Figure 3.2. Four significant points relate to Cheltenham's employment base by the early 1970s. Firstly, although Cheltenham had moved rapidly towards services employment (a national trend), manufacturing industry still accounted for one-third of the workforce even by 1971. This proportion had fallen by 1981, but the sector's employment had actually grown, whilst the proportion was maintained in 1984. In other words, Cheltenham's overall employment growth in the 1970s was partially due to manufacturing performance. Evidently, success was not unequivocally attributed to the locality's attractiveness for insurance companies. According to an analysis comparing Cheltenham's industrial employment growth with national industrial performance, Cheltenham industries either grew faster or declined less sharply than the equivalent industries based elsewhere (Livingstone, 1987a). Part of the explanation is to be found in engineering. While Cheltenham's engineering structure was maintained in the 1970s, mechanical engineering and manufacturing dramatically declined in the UK's traditional manufacturing centres. By 1981, engineering employment was relatively more significant to Cheltenham than it was to Middlesbrough or Liverpool (Cowen, 1987b). This trend, as we shall show below, has continued through the 1980s.

Secondly, the areas of expansion in Cheltenham's manufacturing sector have been dependent on state funding in the form of military contracting, so that Cheltenham forms an integral part of the defence belt stretching from the Severn to the Wash. Since 1979, public expenditure policies have favoured those regions of Britain undertaking certain types of military production (Bateman and Riley, 1987; Lovering, 1985). Such employment has also been increasingly dependent on the procurement of defence contracts from foreign governments (particularly equipment for the American Army and Navy), presenting particular ramifications for the workforce in such localities.

The role of the state also proved to be substantial for slightly

Plate 3.1 One of the many companies of the Dowty empire, at the heart of the locality's extensive subcontracting network in the sphere of defence-related production.

different reasons in the case of Cheltenham's finance and business services sector. Office decentralization policy favoured localities such as Cheltenham and Swindon, which were relatively close to London, offered cheaper land and plentiful labour, and, in Cheltenham's case more than Swindon's, could throw in special local environmental benefits.

The third major point is that the local 'prime movers' other than the state have been multinational organizations. These have included companies where the ownership, control and hence major strategic decision making have been situated outside the locality (e.g., Smiths Industries and Mercantile and General) as well as some whose origins lie inside the locality (e.g., the Dowty Group). However, the scale of a company like Dowty means that its direct relationship to the locality is perhaps more tenuous, in terms of the ability of the local authority to influence employment developments. Almost 21,000 jobs were provided by just five organizations in the area (counting the Dowty Rotol plant of 3000 employees at Staverton, not included under the Cheltenham T.T.W.A. statistics).

The fourth point relates directly to the locality's social structure and cultural ethos. The occupational structure has moved incisively in favour of the professional and managerial groups. Growth in numbers of people working in the Cheltenham area as employers, managers and in intermediate non-manual occupations was faster than in Great Britain in general, reflecting the increasing local significance of scientific, technical and professional occupations. Yet, at the same time, Cheltenham's unemployment figures have been relatively high when compared with localities such as Cambridge or Crawley (both closer to London and with a much higher proportion of high technology electronics engineering). Given Cheltenham's longer-term industrial history, layers of unskilled and casual workers remain unable to obtain regular employment. Cheltenham is not an unequivocally middle class locality. Managers, owners and professionals may have replaced ex-colonials, but Cheltenham has maintained an identifiable if submerged manual working class. The cultural and political ramifications of these features are highlighted towards the latter part of the chapter. In the following section we trace in further detail specific local contemporary developments reflecting the broader processes.

The restructuring of Cheltenham's local economy

By the 1980s, both Dowty and Smiths were in the top one hundred of the 1000 largest UK companies as listed in *The Times*. Dowty employed 15,818 people globally, of whom almost fifty per cent were employed in aerospace and 9000 in the ten local plants (all with the exception of Dowty Rotol falling into the Cheltenham TTWA) manufacturing precision engineering products, fuel injection pumps, hydraulic roof supports for mines, and industrial seals. Despite the apparent diversification, Dowty's military sales of £275 million still constituted more than sixty per cent of their total sales (Labour Research, 1986, p. 8).

London-based Smiths Industries' global employment fell from 15,200 to 11,000 between 1982 and 1986, while its UK workforce declined from 22,000 in 1976 to 9000 in 1986 (Smiths Industries, Annual Reports; Interview with SIAD management) – indications of major restructuring in Smiths' overall operations. SIAD's plant at Bishops Cleeve, with 3200 employees almost exclusively engaged in aerospace and defence production of military measuring instruments, has maintained most of its establishment (whereas the corporation's vehicles division has disappeared). In the 1980s local aircraft and defence engineering contracts have grown, especially from foreign states and international 'tie in' arrangements such as the European Airbus project (Dowty Annual Reports; Smiths Industries Annual Reports; *Gloucestershire Echo*). Orders were at their highest ever levels, including equipment for the Fokker 50 and 100, and the Jaguar aircraft (*Dowty News*; Smiths Industries Annual Report). Buoyant production levels have also tugged in their wake a plethora of small subcontractors, acting as a cushion for the multinationals. Specific information on their subcontracting activities is difficult to obtain, but interviews with management have indicated that the local manufacturers employ at least two hundred and fifty subcontractors nationally and internationally. Many of these are small office or even home-based one person businesses, providing either highly specialized precision engineering services or computer software services; or individual subcontracted programmers working in-plant.

In response to intensified international competition, the locality's major firms have expanded their overseas markets (sixty per cent of turnover) and diversified into a range of increasingly specialized, high-quality products. This trend has appeared alongside three related developments – intensified use of new technologies; organizational rationalization or 'streamlining'; and corporate acquisitions.

Application of CAD (Computer Aided Design) – CAM (Computer Assisted Manufacture) technology has transformed the design process in the local aerospace-related engineering companies. Smiths Industries Aerospace and Defence Systems has replaced hand lathes with automatic tools and spent £4.7 million on a building to house new computer facilities. Between 1976 and 1986 they invested £100–£120 million locally, and set up a micro-processor manufacturing company at Tewkesbury. The general implications for the local workforce were the redundancy of older mechanical engineers in the local labour market.

Company reorganizations of staffing and production methods have become part of the new work climate as efficiency becomes an even higher priority. The corporations have devolved into smaller, immediately identified profit-making units, whilst the system of financial accountability to the parent organization has been simultaneously strengthened. Pressure to keep abreast of innovations has meant the frequent reorganization and intensification of resources and manpower. Total engineering jobs fell by some 2.5 per cent between 1981 and 1984 (Census of Employment), although orders and production increased. Within the Dowty Group, for instance, new graduate managers from across the country, intent on furthering the strategy of modernization and efficiency, have replaced the older, more loyal company employees. Wider global perspectives have led to company acquisitions both in the UK and abroad. Due to shortage of time for nurturing the appropriate new skills locally, the Corporation has bought electronics firms in the south-east, such as Middlesex-based Gresham Lion, producing visual display terminals, transformers and computer graphics (Dowty Annual Report, 1985–86).

Thus, shadowing the mood of euphoria at the list of orders stretching for twenty years ahead, pressures of change have been transforming the engineering work environment, creating a new work culture. But there also exists an awareness, particularly among management and trade union activists, of the consequences of the aircraft industry's volatility. 'If it wasn't for the aerospace industry, Gloucestershire would be an engineering blackspot on a par with Merseyside and the North-East.' (TASS union representative, *Gloucestershire Echo*, 1986.) In the light of such disquiet, the upsurge of Cheltenham's finance and insurance sector has acquired an added importance. In 1981, 9.5 per cent of the Cheltenham travel-to-work area's workforce was employed in insurance, banking and other related activities. The proportion had grown by 1984 to 10.5 per cent, compared with approximately four per cent nationally. Clearly, the 1960s office decentralization policy provided the initial impetus to the employment growth of this sector in Cheltenham throughout the 1970s, at a time when it was the nation's fastest growing employment sector (Boddy et al., 1986). After Eagle Star had set up its headquarters in the town in 1966, the insurance firms that followed in the 1970s were all London-based companies decentralizing their head offices. The largest of these was Mercantile and General Reinsurance, now a subsidiary of the Prudential Corporation, which moved in 1973. Endsleigh Insurance, the National Student Union's wholly owned subsidiary company, moved to Cheltenham in 1972.

A network of insurance firms and brokers developed in the town. By 1984, insurance alone provided 2043 jobs in the Cheltenham travel-to-work area, constituting 2.9 per cent of total employment, double the national average. As with the engineering sector, the prime movers have been multinational corporations, fundamental to the recent development of Cheltenham. Eagle Star's Cheltenham offices, employing 1000 people, act as corporate headquarters for the computer, administrative, personnel and life insurance departments. Due to its precise centralized computer head office functions, the Cheltenham office has sustained office employment. Furthermore, new unit trust schemes have meant a relative buoyancy in the life assurance market, encouraging more recruitment in Cheltenham. In the case of the local Eagle Star operation, this contrasts with its national corporate trend. Even under new ownership by British American Tobacco in 1984, no redundancies have occurred to date, and physical expansions are planned in the town. Whilst this suggests a continuation of local autonomy, household interviews demonstrated that for some the change of ownership has introduced a slight mood of uncertainty for the longer-term future. Mercantile and General, too, have expanded their administrative headquarters since their arrival in 1973. But the ties with London are ambiguous. 'There is real office rivalry,' said a member of the Cheltenham management, 'London is seen as snobby and superior; Cheltenham's looked on as a rustic backwater' (Interview with Mercantile and General Management.)

Indeed, both Eagle Star and Mercantile and General had initially sought speedy access to London as a prime consideration. Most of the insurance and insurance brokers' operations in the area have been founded on an underlying spatial division of labour where 'the business is done in London; the administration in Cheltenham' (although at M&G both business and adminstration have been split). The Cheltenham insurance industry is a product not only of office decentralization policies in the 1960s, but of multinational corporate flexible strategies which have sought to maximize profits and reduce labour costs by the separation of corporated functions on a spatial basis (McNab, 1987). In the process, there has been a further division of labour, which has heightened gender disparities. Some sixty per cent of the workforce in the large insurance firms were female, and forty per cent were male. But on the managerial and professional side, males tended to form the majority of the imported staff, with the females recruited for clerical work from the local labour market. During the 1970s, the female proportion was growing, but this is no longer so. In line with a decentralized organizational strategy, the large Cheltenham companies have also developed local catchment area recruitment and in-house training strategies for moulding company loyalty (rather than industry-wide skills).

The largest single service employer in the Cheltenham area is, however, part of state defence administration and surveillance. Such functions have become an integral characteristic of modern warfare. GCHQ, located on two major sites, has employed some fifteen per cent of the local working population in the 1980s. Placed at the centre of a national and international surveillance network, it has paralleled in services the defence-related manufacturing activities of Dowty and Smiths Industries. A functional arm of NATO, it is integrated with its United States surveillance counterpart, the National Security Agency (NSA). Its early location in Cheltenham also appeared to continue the lineage of the area's colonial and Commonwealth connections extending back via a Foreign Office base during the Second World War to the nineteenth century and the Spa's associations with military personnel returning from India. Contemporary GCHQ outposts still include Cyprus and Hong Kong. Again, GCHQ's roots are in modern wartime operations, namely the intelligence gathering service formerly located in Bedfordshire during the Second World War (West, 1986). Its major SIGINT (signals intelligence) functions and personnel have been closely linked to the American network, entailing regular exchanges of up to three years between Cheltenham GCHQ and Maryland NSA staff. Since the organization's relocation a series of expansions has proved instrumental in influencing the town's physical and social structure, its cultural activities, and the very tone of its politics. 'GCHQ have been a bloody nuisance in this town for years,' we were told by a trade union official who felt that the organization's presence has exerted a dampening influence on labour movement activities. 'The Trades Council has done nothing for years because of them.'

Highly qualified scientists, mathematicians and linguists were recruited from the universities in the 1970s whilst the lower grades and clerical staff have been hired locally (Livingstone, 1987b). GCHQ's impact on the local economy has been estimated to be in excess of £50 million in wages alone (Livingstone, 1987b) and the total annual budget at £500 million (Campbell, 1987). The organization's engineering and computer science divisions have each subcontracted work to outside, private companies. The number of local subcontractors, including Racal on Tewkesbury's industrial estate, manufacturing radio equipment to MoD specification, has grown in the 1980s. Racal, a multinational based in the south-east, has been one of the largest British companies in the cryptographic business, and the close connections with GCHQ were cemented when a senior retired GCHQ officer joined Racal's board in the 1970s (Bamford, 1982, xv).

The locality's internationally known private schools help to define it in the public mind. The Cheltenham Boys'' College has four hundred and fifty pupils, and a staff of two hundred and fifty; with 850 pupils and 450 staff (almost two hundred teachers), Cheltenham Ladies' College is reputed to be the largest of its kind in the world (Packenham, 1971, p. 138). Altogether the town's five independent schools house 2700 pupils with over 1000 boarders (*Independent Schools' Year Book*, 1986) A further six independent schools such as Berkhampstead have catered for the 4–11 years old range, whilst a series of small private language and secretarial colleges has been proliferating through the decade (including a new college for the town's Japanese families).

Over the last ten years, by contrast, revelations have been made concerning the continuing deterioration of Gloucestershire's public sector secondary schools. As though to drive the point home, the Boys' College has invested in a substantial programme of modernization (Cheltenham College Prospectus, 1986). Three particular features have been noteworthy in the restructuring of the private schools. Firstly, links have been forged with localities overseas such as Toronto, Houston and Tokyo, not simply for cultural exchange but for experience in the field of electronics (Interviews at Cheltenham Ladies' College and Cheltenham Boys' College). Nevertheless, the Boys' College's earlier traditions of military education have been continued on a extra-curricular basis (Cheltenham College Brochure, 1986).

But a second feature has been indicative of the more immediate impact of their operations upon the locality per se. Financial viability is pivotal in understanding the private and church sectors of education. The main private schools (Cheltenham Ladies', Cheltenham Boys', Dean Close, St Edward's – a current amalgamation of Whitefriars and Charlton Park) and a church school, the College of St Paul and St Mary, have sold large tracts of land and property through the 1970s and 1980s, rationalizing a multiplicity of academic sites in the process. Much of the land sold since 1970 has been used for high-quality housing developments or for special needs in Cheltenham's fast growing middle class suburbs.

The third feature has been the process of amalgamation in both private and public sectors. The College of St Paul and St Mary represents the amalgamation in 1980 of two church colleges, and is one of the country's last voluntary colleges due to the Labour government's reduction in the numbers of teacher training colleges and students. Its ability to survive has so far been determined by rationalization, land sales and the enhanced provision of adult education courses aimed at the area's middle class population (Principal, College of St Paul and St Mary).

Government education expenditure cutbacks have affected the profile of resourcing in the locality's publicly funded further and higher education colleges, producing course rationalization and, again, an amalgamation of three colleges to form the Gloucestershire College of Arts and Technology in 1980. GLOSCAT's responsiveness to local industrial and finance capital needs has led to a plethora of specialized short courses for GCHQ, insurance and engineering companies, and the creation of information technology centres, in addition to large-scale course provision for the Manpower Services Commission. A key problem has been the difficulty experienced by the local Further Education sectors in matching courses to the multinationals' specific high technology engineering labour demands. 'Any locality like the Cheltenham and Gloucester area that was serious about its manufacturing industry would have made a play for at least polytechnic status in the Sixties' (Interview with Dowty's managing director).

'Higher education: meeting the challenge' (DES, 1986), the white paper entailing tighter methods of controlling higher education funding by central government for the 1990s, has thrown a veil of uncertainty over future provision in the locality (*Gloucestershire Echo*, July 1987), with yet another radical reorganization of the county's post-schooling expected. The disparity between the area's privately oriented interests and public sector educational provision suggests a considerable economic and social polarization, plainly demonstrated not only by the loss of higher-level courses at the LEA college, but also by conditions in Cheltenham's primary and secondary schools, many of which have been poorly maintained (*Gloucestershire Echo*), 1985); the absence of county nursery provision (Gloucestershire County Council, 1980–1986) and the imposition of a secondary reorganization which has failed to provide a fully comprehensive system (Gloucestershire County Council, 1984). 'Morale is rock bottom among teachers in secondary schools. I'll be glad to get out of it' (Interview with secondary school teacher.)

To recap on the key developments, certain features of restructuring have been prominent among the locality's industries. Internationalization has changed the relative importance of the locality within the corporate organization of locally based firms, especially in engineering. Rationalization and corporate flexibility strategies are transforming the types of labour required in the area, while the process of feminization has been notable in the producer service industries and the government services. In the light of the material collected on the four prime movers, we can make certain key points relating to local labour markets as a whole. All of Cheltenham's major employment sectors have been dominated by employers and plants of national significance, with a relatively high level of decision making in the locality. Given their national significance, all four sectors have recruited in the professional and scientific field from national labour markets. Company loyalty has been adopted as an essential expedient by the main firms; teachers and lecturers, too, have tended to 'stay put'. Government agencies have been drawn to the locality for its growing pool of professionals, scientists and administrators, but they have also drawn capital and labour in their wake. Family recruitment for both non-professional and professional jobs has been influential. Because of local employment stability, and the preponderant influx of highly educated families, a continuing mismatch has occurred between qualified female professionals and appropriate full-time employment available, particularly in teaching. Finally, but certainly not least, the other segment of the labour market has comprised a large female clerical labour force mostly denied access to the managerial and intermediate supervisory jobs, and unskilled labour in tourist-related employment and consumer services. Along with the unskilled unemployed, the latter have constituted a growing segment of Cheltenham society, increasingly unable to break into the world of connections and qualifications guarding the sought after jobs. Taken together, such developments have played a distinctive role in shaping Cheltenham's cultural patterns.

Work conditions and change

At least two cultures have described Cheltenham in the 1980s: a 'middle class' and a lesser known working class culture. Dowty and Smiths Industries in the long term fostered a distinctive work culture. Their respective workforces tended to remain with the firm and the locality. For example, a Smiths production manager's father had been an industrial supervisor at Rotol for forty years, and his two younger brothers were also employed at Smiths Industries. 'I've managed to progress there without always pushing.' He was not worried about his own prospects, since like many others, he harboured 'no high aspirations.' But the nature of the work had intensified with modern methods of manufacture and the speed of change in aerospace work: 'Morale's on the down. The employee is now better educated and aware of the uncertainties about future contracts . . . . More nations are involved in contracts such as Tornado.'

Corporate management reorganizations have affected the whole workforce, but in a variety of ways. At Smiths Industries, more decisions were being made in London, notwithstanding the Bishops Cleeve management's efforts to project an aura of continuing local autonomy. 'We've got more of the Thatcherite regime here now,' said a senior union official. 'We're also rudderless. We've no real manager.'

Decentralization within local companies has created problems for the trade unions, which have been relatively strong and well-organized at Smiths Industries due to engineering workers who came from the London region in the 1940s.

'All of a sudden new systems have come. Computer people are replacing the traditional engineers. They're less eager to fight in their type of unions . . . We're more fragmented with the location of the new computer firm out on the Tewksbury Estate' (Interview, union official).

In this respect, the application of new technology on various categories of employees has exerted a set of uneven effects. Michael, 42, an instruments assembler at Smiths (where his father and brother have also worked), had no formal education qualifications and felt the work slipping away with the growth of production by computerization. But within the same firm Alvin, a graduate systems analyst in his twenties, a recent in-migrant to Cheltenham, was employed by Smiths Industries as a subcontractor on a high salary. His wife worked in the area. 'I can name my price. Why be in a union?' he argued.

The shifting technology has unequivocally benefited such an employee. Similarly, Rhona, another recent in-migrant from Scotland, and a computer scientist, was one of a gradually increasing number of graduate female engineers. Single, aged 26, with her own 'starter' home on a modern Cheltenham housing estate, her earnings had risen constantly. The job market was clearly in her favour. She regularly attended job interviews. 'Companies don't leave you alone once you inquire . . . I've already had three 'phone calls from one firm.'

But the intensifying competition for contract work has meant constant and deepening overtime pressures for many Dowty employees though not for their Smiths counterparts, whose stronger union shops have aimed for higher basic pay rates and less overtime. Chris, a middle manager in his thirties, was working seven days a week. 'There was a lull in the early 'eighties, but I'm working as hard now as in the 'seventies. I have to bring my computer files home.' Nevertheless, in terms of job security, there has been a relative absence of immediate anxieties among the full-time engineering workforce of the big companies. And although the local engineering wage levels were distinctly low in the interwar years, recent years have seen improved wage levels comparable with the highest nationally for two basic reasons – the full order books of the big local combines, in search of particular skills and needing to meet tight deadlines; and the organization of the unions at those local plants with a greater composition of postwar in-migrants from the large conurbations. Smiths Industries (Cheltenham) company minimum rates in 1986 were among the country's top thirty of basic grade manual rates (Labour Research, 1986).

Even within the Dowty Group, which has a less militant union tradition, the unions have retarded the rate of new technological application and resisted Japanese-style management innovations such as quality circles: in this case an older workforce, with a long history of craft engineering, has held to basic trade union traditions. Smiths shop stewards on the other hand, have accepted technological changes (Smiths Industries Agreements Handbook, 1984) but displayed wage rate militancy – as during an eight week dispute in 1979

(*Gloucestershire Echo*, 1979). Ironically, a managing director found unions in the north and the south-east easier to cooperate with:

'It's not true about weak unions down here. Our management team puts in sixty or sixty-five hours each week because they're driven by the will to win. But we can't instil it in the workforce. They don't like the outsiders brought in to make the company competitive.'

On the other hand, at Dowty Fuels, with fewer incomers, the unions were weaker. According to a union representative

'There've been more of the cap-touchers from the Cheltenham area at this part of the Group. They've introduced quality circles under another name. But they don't work . . . No one really cooperates.'

Unionization has been relatively weak within the insurance companies in the locality, and indeed some company policies have been explicitly anti-union. Walter had been encouraged by his management to join BIFU 'to counter leftist tendencies in the union'. At Maritime and General there has been a staff liaison committee instead of a union. The workforce is personified rather by a strong loyalty to the individual firm, in cases spanning generations. Vera, as a typist, and John, in middle management, had worked for M&G for more than thirty-five years between them (in both London and Cheltenham). At both Eagle Star and Mercantile and General, there was a distinct feeling of 'a job for life,' since many employees and managers have joined a company direct from school. One young employee's background provided an interesting example of the company connection. His mother, a cousin, two of his aunts and his sister (previously) worked for the firm. Although he had failed his G.C.E. 'A' levels, a prerequisite for recruitment, he had successfully secured a place on a Youth Training Scheme course at the company. On termination of the course, he was offered permanent employment. Apart from job security, loyalty has been strengthened by remuneration packages encompassing a good salary, free restaurants, subsidized mortgages and cheap insurance. For managers, the 'perks' have included a car and private health insurance. But job satisfaction has been more marked among the males. 'There is a definite bias in this company against promoting women,' a Section Assistant Head told us.

An Equal Opportuntunities Commission investigation at Eagle Star (Collinson, 1985) had confirmed such sentiments. Nor has apprehension of change been universally absent. Ray, a graduate recently moved from his company's London office, was able to determine his own work rate 'at the moment. The point is, though, that morale depends on which section you work in, and how quickly computerization is introduced.' A similar caution was observable with respect to recent shifts in corporate ownership. Greg, a manager in his thirties, had arrived from another industry in the south-east in 1983, and within three years had rapidly increased his salary:

> 'But with British American Tobacco's takeover of Eagle Star, there'll be plenty of change in the next few years . . . new blood, and a faster rate of adoption of new technology.'

The work situation in an organization such as GCHQ has differed from those in other sectors, given the overriding factor of the Official Secrets Act and national security restrictions. In local terms, the traditions of the workplace for GCHQ employees and their families are naturally dissimilar from those of the longer standing engineering companies. Many employees of the organization came in the 1970s, with the expansion, and tended to be high-level professional and scientific employees. Although salaries have been relatively poor, in view of the skills demanded, the 1984 management-union tensions have hastened the introduction of a comprehensive regrading system to reflect the specialist nature of the work at GCHQ. But the 1984 events, and the pressures of the state and NSA, produced a distinct effect on workplace morale (Livingstone, 1987b, pp. 30–31). Indeed, household interviews showed that employees deemed this decline in morale to be more serious than any increased workload or introduction of new technology (suggesting a rather different climate to that observed in the local engineering companies). Like the engineering defence multinationals, GCHQ management has become more remote from the workforce – doubtless reflecting the tightening reins of NATO defence policy and US influence. The Conservative government's union ban had embittered personal relations between the workforce and immediate superiors, eroding the image GCHQ initially created of a paternalistic employer with desirable facilities and day release educational opportunities.

Cheltenham: local class and housing market

An older ruling class presence in the immediate environs of Cheltenham has maintained its influence upon the locality's contemporary culture. Indeed, the built environmental features stand as a bridge between an old culture and the new by dint of the recapitulation, in the building preservation policy, of the Victorian Spa ambience and

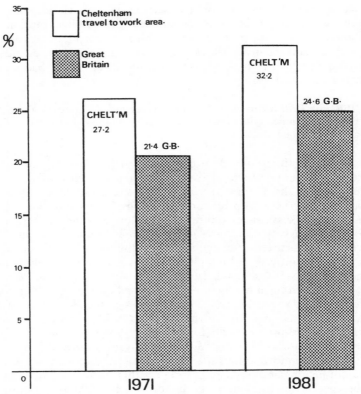

Figure 3.3 Proportion of households where head of household is in professional, managerial or intermediate socio-economic group. Cheltenham travel-to-work area, 1971, 1981. Source: OPCS. Small area statistics (10% sample), 1981.

images (Cowen, 1987c). Yet the changing patterns of industrial capital, the State's strategic defence policies of the 1980s, earlier office relocation policies of the 1960s and 1970s, and the changing social profile of the locality have all modified the conventional Cheltenham ambience. Professional and managerial employees accounted for twenty-seven per cent of total employment in the Cheltenham travel-to-work area at the beginning of the 1970s. They comprised over thirty-two per cent in the 1980s (as against twenty-five per cent in the country as a whole.) This group was growing twice as fast as in Middlesbrough or in south-west Birmingham (OPCS, 1971, 1981). As would be expected, the Cheltenham population is better qualified than Great Britain's population as a whole (7.4 per cent of people in Cheltenham's Policy Area hold a degree or professional qualification, as against 5.6 per cent nationally). Compared with a 1.1 per cent national employment growth, Cheltenham Policy Area's working population grew by 3.4 per cent from 1971 to 1981. Cheltenham registered the lowest rate of unemployment increase between 1972 and 1982 among a series of UK localities (Cooke and Morgan, 1985). Figure 3.3 indicates the proportion of the workforce falling into the managerial, professional and intermediate socioeconomic groups.

Mirroring the more general movements of the professional and managerial groups to the green-field areas of the UK (Fothergill and Gudgin, 1982; Massey, 1984; Hall, 1987), contemporary Cheltenham's marked affluence may be viewed as the other profile of the national economy's structural decline, a key feature in the changing social and spatial division of labour and the fostering of an unemployed underclass in localities such as south-east Liverpool discussed elsewhere in this book.

During the 1970s, the number of persons in the highest categories of occupational groups (owners or executives) multiplied by some thirty-five per cent in the Cheltenham travel-to-work area, and over the same period the professional employees increased in number by some twenty-four per cent (OPCS, 1981). This portrayal of an increasingly evident, higher-income middle class is reinforced by measures of consumer living standards. Almost forty-five per cent of households in the Cheltenham travel-to-work area were without a car in 1971, the proportion falling to 32 per cent (compared with forty per cent in the whole of Great Britain) by 1981. Fewer than 14 per cent of owner occupiers had no car (OPCS, 1971, 1981). On the other hand, as many as seventy per cent of households on the Kirkby public sector housing estates were without private transport (Broadbent, 1985).

One detects, then, a symmetry between the locality's economic structural developments and the local penetration of the current epoch's most dynamic (social) stratum – the 'service class' (Abercrombie and Urry, 1983), recasting but nevertheless guided by the older and previously dominant local ruling class culture. In the changing spatial division of labour in the U.K., localities like Cheltenham have 'elevated the private over the public in cultural and social provision' (Cooke, 1986). This is apparent in the arena of its modern retailing developments, which have married the private individualism of a rising middle class to the style and symbolism of a more leisured conservatism on the one hand, but have progressively excluded working class culture and provision on the other.

The locality's direction in the 1980s is encapsulated by the inner area's property and shopping developments, and the wider Policy

Area's housing. The major scheme, Regent Arcade, has proved to be architecturally noteworthy for the manner in which a modern design has been imposed upon the classical vernacular street plan of old Regency Cheltenham (Herzberg, 1985) and symbolic in its style and facilities, reflecting the consumerist aspirations of a new professional and white-collar service class (Cowen, 1987a). Indeed, the protests of the pure preservationists to keep the original interiors, the Plough Hotel and the surrounding old buildings, were clearly muted by the power of property capital. In the process, the former retailing superiority of the city of Gloucester over Cheltenham has been reversed. Cheltenham has now attracted the multiple outlets such as British Home Stores and C&A, in response to the consumer requirements of the burgeoning Cheltenham suburban estate occupants. Yet, in contrast to what happened in Gloucester, the prime thrust has been the expensive and fashionable pockets of the market, consumer schemes intended to match the prestige images projected by the producer service sector. Such a trend has been conspicuous in the methodical rehabilitation and upgrading of the Montpellier shopping neighbourhood, and the reconversion of the Promenade's major post office building into exclusive dress shops and restaurants for wealthy residents and tourists.

Cheltenham Borough's policies of the current decade have in turn produced certain indirect impacts upon industrial trends in the wider commuting area. Industry has been consciously deflected, while the modernized Spa image and prestige shopping facilities have been prioritized for the development of a tourist economy and for servicing the 'service class'.

High technology engineering firms drawn to the area and the M5 since 1980 have located on industrial estates which lie partly within Cheltenham (Cowen, 1987b). As opposed to the conservation-based Cheltenham Borough, Tewkesbury planners have been implementing a robust set of economic initiatives to entice high technology industry and multiple private housing developments of a similar character and on a similar scale to those in the Swindon area. Unlike Thanet, the geographical centrality of the area and its environmental attractiveness have been major marketing factors. Paradoxically, this is extending pressures on the neighbouring Green Belt and Area of Outstanding Natural Beauty in the Cotswolds, traditionally a strong incentive for incoming labour. Furthermore, it is producing political controversy over the local government boundary proposals and their effect on potential rateable income for the rival local authorities (Cheltenham Source, 20 Nov., 1986).

A further representation of the privatizing economy can be seen in

Plate 3.2 Regent Arcade, symbol of middle-class Cheltenham. The first contemporary mall in the locality, built along classical lines (with a touch of the postmodern). More such arcades are planned.

the Cheltenham housing and property markets. The housing market's key growth area has been in private housing, and the extent of owner occupation in Cheltenham's travel-to-work area, sixty-five per cent, stands well above the national average of fifty-six per cent. Furthermore, owner occupation was far more significant in 1981 than it had been a decade earlier. The spread of private housing estates to

Table 3.2 *Percentage change, housing tenure, number of households, Cheltenham TTWA, 1971–1981*

	1971	1981	% Change 1971–1981
Owner Occupation	25059	34434	37.4
Council	10201	11247	10.3
Others	6449	6586	2.1
All	41709	52267	25.3

Source: OPCS, 1971, 1981

suburban Cheltenham and the adjoining parishes came in response to the excess demand for higher quality housing in the locality from in-migrants. By 1985, the Gloucestershire Structure Plan's housing allocations for the Cheltenham Policy Area up to 1996 had already been accounted for. The majority of the locality's population growth and new housing developments between 1976 and 1984 took place in the Tewkesbury Borough parishes of Bishops Cleeve, Swindon Village, Prestbury, Leckhampton, Woodmancote and Up Hatherley (Gloucestershire County Council, 1985). Large private housebuilders such as Bovis are focusing on a series of urban villages of newly built detached houses for their current expansion programmes, facilitated by substantial release of land by the Tewkesbury Authority. Residential and industrial developments have been increasingly deflected to suburban and urban-rural fringe localities.

More recent pressures on Cheltenham's housing market have come fom the importation of London house price inflation into the Cheltenham TTWA. At the more expensive end of the market, notably in the surrounding Cotswold villages, house prices in excess of £150,000 have become typical (*Gloucestershire Echo*, Weekly Housing Supplement, 1987). This has reflected the growing purchase of expensive residences by the London area's managerial and professional groups, for whom the Cheltenham-Tewkesbury prices seem relatively cheap after the recent upsurge in London house prices (Pennycote, 1986; Hamnett, 1987).

Another important feature in the local private housing sector has been the boom in two-bedroom starter homes for single people in their twenties, a boom fuelled by the large insurance companies based locally and their policy of advancing mortgages to their employees mainly for the purchase of new houses. Privatization in the Cheltenham area has been furthered by the rapid development of both sheltered housing schemes and private nursing homes for the elderly in Cheltenham's urban environs. This is similar to the expansion of the private sector in care for the elderly in Thanet (Buck et al., in

Chapter 5.) With respect to the sheltered housing projects, estimates suggest that at least fifty per cent of the purchasers have come from other parts of the country (Information from Pegasus Developments). A 1986 Cheltenham Borough Council report on housing for the elderly lists nine current and recently opened shared ownership schemes providing for 264 flats, twenty-four houses and two bungalows, built by specialist developers such as McArthy and Stone, all constructed since the beginning of the decade. In addition, there were thirty-seven private and voluntary homes for the elderly, all but one registered with Gloucestershire County's Social Services Department, housing another 728 persons. Ironically, the planners have expressed an underlying concern that the borough's demographic structure will once more become overly skewed towards the retired – the nineteenth century spa revisited! – with the consequent burden on public sector social services already under strain.

But the private modern housing developments have been paralleled by the activities of the local authority, whose policies have in no small measure helped to steer the locality's current progress. Rehabilitation expenditure incurred by the local council in the 1980s has produced effects which have been of undeniable advantage to the higher income earners – largely as a result of the upgrading of the inner area housing stock and the consequent conversions into upmarket apartments for single professionals. Grants to owner occupiers living in Regency properties within designated neighbourhoods have meant a subsidy to private housing tenure. Through the 1980s the local authority's grant spending has averaged approximately £89,000 per head, compared with £55,000 in Gloucester and £62,000 in England as a whole (Interview, Cheltenham Borough Council deputy treasurer, 1986). But it has also contributed to an increase in the general level of house prices, and has made the Regency housing and apartment market distinctly upper income and middle class. 'What the private housing market is doing now is undoubtedly being influenced by the local authority' (Interview, secretary of Cheltenham real estate agents).

During this period not only did the number of real estate agents in the town rise to sixty from fewer than thirty, but the council housing stock dropped in excess of 9 per cent from 1982 to 1986, while more than fifteen per cent of the stock was sold (CBC, 1986). The local Shelter group claimed in 1986 that at least 650 people were having to live in bed and breakfast accommodation, whilst scant finance was available for non-statutory assistance in this area. Cheltenham Council's total support of advice services, less than £30,000 for the 1987/88 financial year, is well beneath the national norm (Shelter, 1986).

Cheltenham ambience, local policies and the built environment

Local planners, increasingly concerned with marketing tourism in a competitive 'municipal' climate, have built their present campaign around the features of the town's building stock and the overall environment. 'What the town offers is represented as much by its activities and "feel" as by its physical attributes and facilities.' (Cheltenham Borough Council, 1986.) Both capital and labour have migrated to the locality in recent decades, their presence attracting further economic activities in their wake. Given that the Labour government's office relocation policy of the 1960s was aimed at pressurizing offices and firms to move away from London, Cheltenham's particular location just beyond the restricted zone was, like Swindon's, an advantage. Office land development was facilitated into the 1970s. Cheltenham Council concentrated on luring the high prestige companies and offered choice locations, new buildings and rates holidays. It then reorientated its strategy towards the Regency stock as a resource for converting into prestige office buildings. A sequence of council policies enticed insurance companies such as Mercantile and General, government departments and agencies, and organizations such as Kraft Foods and Gulf Oil to set up their head offices in Cheltenham. Early beneficiaries of the direct inducements included Eagle Star (who produced Cheltenham's only skyscraper, against the council's wish), drawn by the concessions on rates and permissive planning approvals (Hurley, 1979).

Officials were anxious to receive company management representatives. The Countryside Commission, which vacated its London location in 1970, was predominantly interested in access to the National Parks rather than proximity to the capital, but environment and the architecture ranked high in the discussions with the government's Property Services Agency. But other considerations counted, too – the local offer of a noted Regency heritage edifice which was afterwards converted, or the fact that the Countryside Commission's chairman happened to reside in Fairford, in the Cotswolds area. 'Clearly, the character of Cheltenham wins out over Rugby, if other things are equal.' (Interview with current chairman, Countryside Commission.) The Universities Central Council on Admissions (UCCA) relocated in Cheltenham during the same period, and later in the 1980s its public sector counterpart, the Polytechnics Central Admission System (PCAS), also moved from London to the locality, Initially the latter was motivated by proximity to UCCA, although later it was more impressed by 'the benefit of a local pool of well-educated temporary female staff' (Interview with PCAS chief executive).

During the early 1970s the local state's Regency building conservation policy was subsumed by its office development policy and the town's financial and property interests. Accordingly, because of the enthusiasm to market the town's historical image, finance was forthcoming for the conservation of building exteriors through the conservation fund set up by town and county in 1966, although at least half the conservation grants have come from central government. At first, the policy applied to Regency buildings for conversion to offices, but was then widened to encompass the Regency housing stock per se, perceived as an indispensable motif in the locality's public image and corporate status. As a consequence of protest by the conservationist lobbies in the town, the office development policy was finally curtailed, to be replaced by a policy of comprehensive rehabilitation of the Regency housing stock and the institution of special conservation areas which embraced residential neighbourhoods and exclusive shopping areas. This approach has operated in tandem with the city centre retailing policies of the 1980s, the present tourist strategy and a predisposition towards producer services (CBC, 1966). The prevailing borough plan stresses the conservation emphasis of Cheltenham's architectural image (CBC, 1985). In fact, the town's central district has been designated a conservation area. Two thousand buildings are listed as possessing special architectural or historical merit, in addition to those of intrinsic local interest. By the mid-1980s, grants worth more than £9 million had been steered into the locality's private Regency housing stock. The restoration of Regency properties accounts for £395,000 per annum. Estimated expenditure on improvement grants for 1987 was £2.5 million (CBC, 1986). Influenced by the prestige of the grand Regency mansions refurbished under the programme, financial houses and building societies like the Chelsea Building Society have relocated their head offices in such properties. The Chelsea has also undertaken a £2 million building and equipment expansion on adjacent land, to double its head office staff by four hundred by the 1990s (*Gloucestershire Echo*, October 1987). The environment has not only affected decisions based on capital's demand for land and built resources. It has also been held in high regard by labour, especially among those with more marketable skills. Our research into the impact of environment has in fact produced pointed examples of careerism being traded for housing or the less tangible ambience. A head teacher described Cheltenham as 'the graveyard of ambition'. A deteriorating work situation in the town's public sector schools has not resulted in strong migration to other localities. Turnover has been low. Teachers have opted for the relatively more comfortable existence in a place 'as nice as anywhere.' 'Many teachers have been here too long,' said a Cheltenham head teacher.

If anything, this has had singular repercussions upon females. With a combination of few vacancies and the wave of in-migrant professional families since the early 1970s, Cheltenham has become a locality of educated, well-qualified female teachers and college lecturers, unable to obtain full-time employment but forming a pool of 'supply' and part-time teachers (Cheltenham Job Centre).

Environmental trade-off, however, was not peculiar to teachers. Explained the manager of a local advertising agency:

'Cheltenham has enticed advertising people more for the environment than the money.... Coming from London, they've been able to purchase the Cotswold Stone houses or the town centre Regency properties....But they aren't the really ambitious high flyers. They're still the guys that the London agencies call "second rangers".'

Jerry, an Eagle Star trainee, found that the workplace tensions posed no special problems 'so long as the living environment is stable . . . . The Regency environment conveys an atmosphere of stability. Cheltenham is a nice place to work.'

Origins seemed to have a bearing on perceptions of the environment. High grade GCHQ professionals from London found the town's architectural heritage insufficient to compensate for a lack of social sophistication. Some of the native skilled engineers, however, saw Cheltenham differently. 'It was a town built for the rich and we benefit', was the view of Ron, a deputy convener, whilst even Nick, a trade unionist who'd been blacklisted by the Engineering Confederation admitted: 'I wouldn't live anywhere else . . . . I like the beautiful old buildings. They were built by the working class.'

Whilst the locality's architecture has been a crucial strand in the maintenance of Cheltenham's public ambience, so too have the arts. The Annual Festival of Music was inaugurated immediately following the Second World War, and the Festival of Literature began in 1948, appealing to an intelligentsia informed by the world view of Cheltenham's public schools. Changes in the Literature Festival began in the 1960s, reflecting more general transformation in the arts throughout Western capitalism. The weakening of the local authority's parochialism was more apparent in the realm of music than in artistic culture per se. 'The Music Festival has always been the golden boy here for the local council', declared a director of the Literature Festival, 'most members of the council haven't the slightest interest in literature.' Despite this predisposition, the size of the Literature Festival grew between 1975 and 1986, during which time it expanded

Plate 3.3 Cheltenham's Municipal Buildings, centrepiece of the spa's historical Regency promenade, where Tory councils dominated the area's politics for decades.

from a weekly to a fortnightly programme of events, attracting 12,000 people in 1986 as opposed to 2000 in 1975. This was due partly to an enhanced awareness of tourist economic benefits for the locality, and partly to the restructuring of the locality's workforce and the massive increase of professionals from the 1970s (CBC, 1986).

At the end of the 1970s, Allen Ginsberg, the American beat poet, appeared at the Festival. By the early 1980s, poets Seamus Heaney and Ted Hughes filled the Cheltenham Town Hall with Cotswold school parties. John Cooper Clarke, an unconventional left-wing poet, also filled the same auditorium with local punk youths. Such audiences have alerted the organizers to the presence of a social subculture previously isolated and ignored. Melvyn Bragg, the TV arts commentator, found the Cheltenham Festival ambience unique and appealing, reflecting a chemistry between reader and audience. 'Small university towns are ideal for that . . . . Cheltenham becomes a literary university during that fortnight.' (Bragg, 1985, p. 58.)

Yet unlike other small festival towns such as York and Lancaster, Cheltenham is not a university town. One again encounters the paradox of an educational nexus, tied to global and local economies, which has fostered an essentially minority culture, without a university. Bragg's misconception may perhaps be explained in terms of the ambience communicated by the town's intellectual and artistic symbols and its manifest public school culture. Most household interviewees recognised the significance of festival culture for Cheltenham, but had no personal involvement or interest. Festival audience profile statistics have provided a useful measure of the relatively low 'local' involvement. Although some forty percent of the Literature Festival audience were borough residents, the majority of the Music Festival audience travelled in from the Cotswold villages, where many of the professional families have moved, and from the West Midlands conurbation (Cheltenham Borough Council, 1986). Dave Dowse, director of the Fringe Music Festival, felt that the local authority had been blinded by the prestige of the International Music Festival.

'Five years ago there were ninety rock groups performing at the local arts centre. The council strangled the scene. It didn't fit in with the image . . . . It's still a localized place, socially and politically'.

Just as its artistic culture has remained essentially conservative, so the politics of the locality has been marked by Cheltenham's persistent conservative past, which at crucial moments has been inseparable from the conservation and recreation of environmental imagery.

Politics, conservatives and conservation

The idea of natural and architectural heritage has been central to the cultural symbolism of conservatism in the locality, with an evocation of the predominantly green images of Deep England (Wright, 1985) enlisted for the stimulation of patriotism and nationalism as recently as the early 1980s. In this respect, *This England* a national heritage magazine published from Cheltenham, has been described as projecting a colonial ambience which bears testimony to 'the cultural psychosis of modern English/British nationalism' (Wright, 1985, p. 91). An elitist culture has unmistakably prevailed over the locality's politics.

Conservative interests have proved dominant up to the turn of the 1980s. The Cotswold area village culture has communicated itself not just as an ideology of anti-urbanism but as the proclamation of an English nationalism through the medium of rural environment and village life (Newby, 1987). This predisposition towards preservation has also been interpreted as a trafficking in history by a ruling class which views rurality as the national, imperialist saviour of the past ranged against the brutal industrial values of the present (Wright,

1985). The same interpretation sees the Cotswolds villages as a continuing focus for the English middle class dream (Wiener, 1985).

In 1982 Prime Minister Thatcher suggestively chose to deliver a seminal speech during the aftermath of the Falklands War on the site of the Cheltenham Racecourse, in a locality imbued with state surveillance functions, defence manufacturing and public schools. The occasion was fundamentally an invocation of those symbols and memorabilia of imperialism by which she rebuked the doubters who: '. . . had their secret fears . . . Britain was no longer the nation that had built an Empire and ruled a quarter of the world. Well they were wrong.' (Barnett, 1982, pp. 149–153.) Yet, as observed at the beginning of the chapter, Victorian conservatism has been tempered by other traditions in the locality, so that electoral politics in the borough has not always replicated the politics of the shire. A tradition Liberal seat until 1910, Cheltenham has retained its liberal and independent strains despite the Conservatives' electoral dominance for most of this century. The current Conservative MP for Cheltenham, Charles Irving, is a local hotelier and former mayor who has independently campaigned on specifically local issues and remained until 1986 as local councillor, chairing various committees including the county's social services committees. Nevertheless, environmental values have not been articulated exclusively in conservative terms. As Wiener (1985) suggests, anti-industrialism has constituted a common reference point for the English ruling classes and a strain of English socialism. Anti-materialism boasts a tradition in the area dating back to the cooperative and socialist movements inspired by William Morris, whilst the more contemporary environmental movement since the 1960s has included rural cooperative, Friends of the Earth, the Green Party, and the drug culture (Newby, 1987).

Key political issues in the locality reflect the perception of an exacerbated powerlessness to exert any sustained impact on national processes. Accordingly, politics as an activity has been interpreted in local terms. Town planning issues, mainly concerned with conservation versus development, have dominated the local political agenda for a long time.

Yet differences over the theme have expressed themselves through consensus politics across the decades in which the Conservatives exercised total control in both county and district. Thus the contained consensus around planning has tended to reflect the locality's conservatism. Divisions have cut across party lines or have prevailed more at the local authority administrative level, between officers who have been 'conditioned by years of Tory rule' (Interview, Labour councillor).

In the 1960s, the Gloucestershire County strategy for comprehensive redevelopment of the inner area advocating demolition of most of the Regency buildings in the fashion of city planner Colin Buchanan (1963), saw the Conservatives divided, with the Liberals in favour of conservation (Hurley, 1979, p. 164) and Labour generally advocating the plan. The conservation lobby gravitated around the Cheltenham Society (now known as the Civic Society). It included professionals, managers and small property owners. The interests of the last group were successfully represented in that conservation was secured. Success for a conservation strategy over redevelopment suggested that professional and managerial interests, especially in the finance sector, could still be accommodated by office development policies which involved the use of refurbished older properties.

Although conservatives as a grouping have not always acted monolithically, conservatism's longevity has diluted the power of non-conservative pressure groups to effect political change. Attempts by the local Labour Party, labour movement and student campaigners to instigate a mass mobilization for the expansion of council housing in the face of growing homelessness in the 1960 and 1970s experienced little success. The notable example of the local authority house building programme in the 1950s, discussed earlier, had been carried out in response firstly to the immediate requirements of industrial capital when the Conservatives wanted manufacturing industry to build up the locality's economic base during the interwar years, and secondly to the national state's strategic requirements (hence the rendering of state financial backing). Since the 1960s and the diversification into producer services, the most powerful political forces have followed policies according with professional middle class images and expressed in prestigious development schemes.

With an altered industrial structure in the finance capital and manufacturing sectors, fostering a specialized white-collar service class, electoral political changes have served to dislodge the Conservative stranglehold on both the Gloucestershire County Council and the Cheltenham Borough Council in favour of the Liberal Alliance, who have held political power in Cheltenham since 1984. The rise of the Liberal Alliance has been most pronounced at local elections. It failed to defeat the Conservative candidate in the June 1987 General Election, despite the massive transfer of votes to the Liberals when Richard Holme, a former Liberal Party Chairman, failed to unseat the incumbent Charles Irving. Between 1977 and 1985 the Liberals increased their share in County elections from sixteen per cent to forty per cent of the vote, reflecting a large swing away from the Conservatives. For the first time the Tories no longer controlled Gloucestershire politically. The Liberals increased their vote from nineteen per cent in 1976 to forty-two per cent in 1986, to gain control of the borough Council (*Gloucestershire Echo*, 1976–1986). The local press interpreted the 1985 county results as the end of an era:

'The elections have sounded the retreat for the so-called 'Cotswold Cavalry' – the aristocratic ladies and gentlemen from Cotswold manor houses who have made key decisions at Shire Hall for decades (*Citizen*, 1985).

Company directors, farmers, businessmen and the retired gave way to professionals, employees, middle managers, housewives and the self-employed (Carter, 1986, p. 22). Daughters and sons of the formerly solid Labour supporters from working class families employed in engineering works were no longer restrained by the same political loyalties. But although the Liberals have replaced the Tories the conservative legacy remains resilient. Labour representatives and campaigning pressure groups do not see the new liberalism as a change in substance. A Labour councillor claimed:

'The Liberals tend not to have altered things much. Let's face it, the officers have been 'conditioned' by the long spell of Tory rule. They've tended to block the 'new boys.'

The Liberals, although they have made declarations concerning jobs, housing and industry, have not effectively challenged the conservation policy. 'As a politician I am in favour of the Green Belt and new development for jobs and homes,' was the position of the Liberal vice-chairman of the borough's Development Committee. The Liberals also support the strong emphasis on the Regency town centre image and the drive to attract tourists. Funding allotted for employment initiatives is still low (£11,500 in 1987/88 as against £150,000 per annum granted for the restoration of Regency properties (CBC, 1987). But the funding has gradually increased since the Liberals have begun to reorientate the District Treasury's spending patterns, reflecting their declared community concerns such as a sports complex and a flood prevention scheme (Cheltenham District Treasury, 1986).

On the other hand, the internationalizing processes of the manufacturing multinationals in the area suggest that the locally based industry has been decreasingly subject to influence by the local state or local politics – the indigenous firm has outgrown the locality. Dowty's operations in the Cheltenham hinterland span four local authorities but, declared a Dowty managing director,

'None of them has provided the kind of infrastructure an organization like ours requires . . . Cheltenham seems to see itself as a service town. The Cheltenham/Gloucester area as a whole seems to see itself as immune from the outside world.'

Historical contrasts with earlier relationships between local authority and private manufacturing company are sharp:

'Dowty Rotol has little to do with the locality . . . It pays a lot of taxes and employs a lot of people . . . But it sees itelf as an international firm located in Britain, rather than a local firm situated in Cheltenham.' (Interview, Managing Director).

The 1980s have also been a period in which the profile of political activists has changed. In all the political parties, a large proportion of the activists have tended to be women – mainly teachers, housewives or unemployed. Activists have frequently been in-migrants from other regions of the country. The shifting character of the electoral support has been duplicated in the membership of the council. Since 1979 fewer self-employed and retired people have been elected, and more social workers, solicitors and service sector company employees. The local Labour Party no longer has a gas fitter on the council, but is instead represented by a teacher, a retired civil servant and a retired commercial artist.

The Liberals have argued that their style of community politics, imported with the incoming professionals, has suited the non-party spirit of local Cheltenham politics. Yet despite this claim, and the involvement of female activists, there has been no long-term radical collective movement or mass political campaign. 'In fact,' said an Unemployment Centre organizer, 'community strength is very low in Cheltenham. It's the individualistic and competitive attitudes that dominate.'

Poor basic provision in areas of the social services and the depletion of the council housing stock have worsened the local housing situation for many young people. The situation has been aggravated by housing inflation and the council's strong conservation policy which has upgraded the Regency housing stock. Six hundred people were living in bed and breakfast accommodation in 1986. But the collective response has been sparse. 'There's no history of tenants' associations here, for example.' (Interview, Shelter Advice Centre.) 'The oppressed feel they are in a minority. They won't put up a real fight.' (Interview, Centre For Unemployed.)

The absence of a viable women's movement may be explained partially by the absence of a university or polytechnic in the locality. Again, the contrast with Lancaster is instructive. The university town has witnessed not only a long-term women's movement but a tradition of extra-parliamentary socialist politics. 'Grassroots' politics in Cheltenham during the past fifteen years, while not entirely dormant, has been sporadic, usually in concert with general national trends, and more consciously related to students and unemployed youth than to the area's formal Labourist trade union bodies. For instance, the beginning of the 1970s saw the development of a core of left-wing students and staff at the College of Art organized around a series of radical issues concerning art education. This nucleus also provided the momentum behind the Lansdown Tenants' Association and the Cheltenham Housing Campaign. The 1974 campaign was for tenants' direct participation in the joint council–Guinness Trust rehabilitation of the neighbourhood's classical housing (Hurley, 1979, Chapter 6). Significantly, in the light of the economic, political and social changes in the locality discussed above, this was the last instance of a community housing struggle. The only other notable waves of non-electoral political protest were the swift growth of both the Anti-Nazi League and the Campaign For Nuclear Disarmament locally. Again, both were products of a national movement, and both drew their activists from the local colleges and the educated professionals. Whilst the ANL disappeared with its national movement, the local CND has survived with the continuation of its national campaigning base. The latter's current activists still constitute a small hard-core of campaigners on the range of 'liberal' and left-orientated issues. Politically, they have belonged to the Labour Party, the Communist Party (now dying with its older members who came in the 1930s) and Socialist Workers Party, and to a lesser extent the supporters of 'green' politics and the politically non-aligned. Most of them arrived in Cheltenham within the past fifteen years, with the exception of the trade union activists who came to the engineering companies in the 1940s.

Unlike these political campaigns the Campaign for Comprehensive Education, although it has held public meetings, has tended to concentrate in the formal electoral political lobby channels, and has been closely tied to the local Labour Party. Hence, its formation in the 1980s did not give rise to a mass mobilization of opposition to the Secretary of State for Education's anti-Comprehensive scheme, which has now been implemented. But this issue is one further instance of how, despite the internal 'liberalization' of the local councils, the stronger powers of the nation state and national Conservative policies can prove to be an overriding factor. Yet in spite of the significance of the nation state and the impinging of national political issues upon the locality, people in general and the Council have perceived the conservation–development issue to be the major political matter. Such a localized attitude was indeed summed up by the widespread refusal to see the GCHQ dispute as a local issue.

Yet if one series of incidents more than any other has lodged Cheltenham in the public mind in the 1980s, it was the Conservative

Plate 3.4 Where the issue of political rights clashed with the state. Plenty of international media coverage, but less impact on the locality's inhabitants.

government's ban on trade unions at GCHQ on 25 January 1984, which unexpectedly galvanized large sections of the white-collar trade union movement. This was part of the Conservative policy of taming what was perceived as an overly powerful national trade union movement, a strategy which entailed direct state intervention in a number of disputes.

The GCHQ trade unions dispute is central to this analysis of the Cheltenham locality because it captured a crucial moment in which the local definition of rights came into conflict with global defence policy and, in particular, increasing NATO–United States pressure on issues of international surveillance. Rights of trade unionization were attacked at a security base in an apparently conservative environment which had long projected an old Regency image, in the belief that organized workers would prove vulnerable on such inauspicious terrain. In the event the government misjudged its opposition, and for many weeks the town was the scene of intense trade union and political activity. A series of televised mass meetings were held at the Pittville Pump Room, historical symbol of the Spa, leafletted by sellers of 'Militant', Morning Star and Socialist Worker. Demonstrations took place in the town centre, and have continued on an annual basis.

Conclusions

Manufacturing industry, as we saw, did not arrive on the Cheltenham scene until the twentieth century, to be greeted by the surviving remnants of a colonial ruling class culture, the institutional passivity of private education and a tradition of domestic service for the working classes. Cheltenham in change bears all the marks of a cultural hybrid, partially driven by the more ubiquitous economic organizational forms and pressures of modern capital accumulation, yet shaped by the accumulated economic, political and cultural layers of an older class structure.

Just as the character of Victorian Cheltenham's economic and social structure was powerfully governed by the locality's ties with British colonialism and a state aristocracy (both the milieu and the officer class of colonial life being reproduced back in 'the mother country') so its structure in this century has been tethered to national state strategy, international treaties and war. Manufacturing flourished during wartime. Military requirements for precision instruments moulded the production profile and local labour markets, and indeed succoured a paternalistic local engineering work culture. But in mid-century, administrative white-collar employment was developed by the Civil Service and Foreign Office.

Each strand was instrumental in contributing to Cheltenham's major layers of investment. The importance of blue-collar skilled engineers after the Second World War supplemented the skilled specialisms of the area's aerospace engineering. On the other hand, the arrival of GCHQ meant the beginnings of a new administrative professional service class, a modern meritocracy in place of nineteenth century ex-colony administrators. It also represented the first step in what has become a long-term postwar restructuring of the Cheltenham economy. Government was again a significant ingredient during the next major step in restructuring, at the end of the 1960s, when the local decision makers seized the opportunity provided by state decentralization policies to diversify further the locality's industrial structure away from engineering. It was propitious that the diversification moved together with a UK economy which was rapidly shifting towards service activites and out of traditional manufacturing. Cheltenham thus became an integral part of the changing spatial division of labour, a southern recipient of jobs geared to growth and high prestige industrial activity, whilst the heavy manufacturing jobs in the northern cities and central London simply collapsed.

The combination of national state investment, government policies, and international restructuring constitutes a convincing set of explanations behind Cheltenham's particular course. Yet we need to bring more sharply into focus the specific local qualities. The earlier state investment in administrative employment and the municipal housing ventures, coupled with the locality's private building resources and the related cultural ambience, offered a particularly attractive package of labour market skills, environmental quality and social prestige – in other words, a highly marketable image. The dialectic between the global and the local has fashioned the developments in Cheltenham over the past fifteen years.

From the later 1960s until the current period, the influx of private financial services and government administrative organizations, as well as the GCHQ expansion in the 1970s, have greatly enhanced the locality's occupational structure by increasing the proportions of employers, professionals and managers. This trend has paralleled the implementation of industrial management's internal restructuring strategies such as production and organizational flexibility. A sub-contracting network has been nurtured by the conglomerates, requiring widely educated personnel on the one hand, and highly specialized analysts on the other. National labour markets are becoming increasingly significant for the local engineering companies with the globalization of product markets and corporate ownership. In turn, this has facilitated a clash of work cultures between the traditionally trained blue-collar worker and the in-migrant professionalized labour force – a specific example of a continuing but also worsening impact of restructuring: the process of social polarization. (Although relative wealth may have been reduced since the turn of the century, new inequalities between the professional middle classes and the unskilled or unemployed are increasing). Local images and the marketing of environmental resources within rapidly shifting national and global economic developments have been integral to the whole restructuring process. The specific ambience of Cheltenham's ruling-class history, its Regency architecture and festival culture, has been subject to conscious manipulation in the attraction of commercial companies and state agencies which have themselves imported professional labour, apart from using a large female clerical labour force. Manipulation of the built environment has been particularly pronounced, oriented towards the needs of finance capital through the maintenance and refurbishment of buildings and prestige public symbols. In this respect, the policy direction has been guided by the locality's conservative and conservation-oriented political forces, tapping the historical mainsprings of an English anti-industrialism characteristic of the Cotswold countryside and built heritage. With the deflection of industrial capital into the Tewkesbury industrial estates, the Cheltenham Borough has chosen property capital as the mainstay of the economy in the 1980s. The strategy has been facilitated by the strengthening of local authority conservation policy and continuous central government funding for the improvement of Regency buildings. Against this policy background, the in-migration of white-collar and professionally-oriented families for employment in the local companies has instigated an era of burgeoning private suburban housing, reflecting the overall privatization policies of Thatcherism in the 1980s.

Economic and social change in Cheltenham, although greatly affected by national restructuring within British capitalism, has displayed not inconsiderable local effects. The role of the built environment, as manipulated by the local state, has been particularly pronounced in that it has shaped the prestige social images of Cheltenham and maintained the ambience of a leisured Victorian town centre. Yet, at the same time, local policies and locality features hold diminished importance for the increasingly globalized operations of the multinational engineering corporations, who nevertheless provide a significant proportion of Cheltenham's employment.

The results of restructuring in discrete localities may be various and diverse, in that they relate to specific and historically unique qualities of the locality. Wheras national and international restructuring processes have generated new rounds of investment in localities, a related process has also occurred, whereby investment is channelled towards a type of locality possessing historically based characteristics not simply economic in origin. Such local historical processes are not only the results of developments and processes generated outside, but also encompass the accumulated effects of local class cultures, local state priorities and the ideologies of local groups.

Acknowledgements

We would like to thank all who participated in interviews during the research for this chapter, and acknowledge the special assistance and information offered by companies, organizations and individuals including Colin Woods, Personnel Manager, Dowty Fuels; Don Smith, Personnel Executive, Smiths

Industries Aerospace and Defence Systems; Mike Marshall, Maritime and General Reinsurance; Tony Roberts, Gloucestershire County Council Planning Department; David Hunt, formerly Chief Forward Planning Officer, Cheltenham Borough Council; Mr Carol Bennett, GlosCat; Anna Dekker, Centre For Environmental Education, GlosCat; and Gerry Metcalf, GlosCat. The research was funded by ESRC, grant No. Do 4250012 under the 'Changing Urban and Regional System' Research Programme.
Photographs: Ian Livingstone.
Graphics: Frank Bushell.

References

Abercrombie, N. and Urry, J. (1983) *Capital, Labour and the Middle Class*, London: Allen & Unwin.

Bamford, J. (1982) *The Puzzle Palace*, London: Sidgwick and Jackson.

Barnett, A. (1982) *Iron Britannia*, London: Allison and Busby.

Bateman, M., and Riley, R. (ed.) (1987) *The Geography of Defence*, London: Croom Helm.

Boddy, M., Lovering, J. and Bassett, K. (1986) *Sunbelt City?*, Oxford: Oxford University Press.

Blake, S. and Beacham, R. (1982) *The Book of Cheltenham*, Northampton: Barracuda Books Ltd.

Bragg, M. (1985) 'Hancox Half-Hour', *Punch*, 30 October.

Broadbent, A. (1985) 'Estates of another realm'. *New Society*, 14 June.

Buchanan, C. (1963) *Traffic in Towns*, London: HMSO.

Campbell, D. (1987) *New Statesman*, 22 January.

Cheltenham Borough Council (1985) *Cheltenham Borough Local Plan*, Cheltenham: CBC.

Cheltenham Borough Council (1986a) *Housing: Monitoring of Progress and Policy 1985–86*, Cheltenham: CBC.

Cheltenham Borough Council (1986b) *Regency Cheltenham Spa: Garden Town of England: Centre For the Cotswolds: A Strategy For Tourism*, Cheltenham: CBC.

Cheltenham Borough Council (1986c) *Festival Audience Survey*, Cheltenham: CBC.

Collinson, D. (1985) *Equal Opportunities Survey: A Preliminary Report*, Manchester: Department of Management Sciences, University of Manchester Institute of Science and Technology.

Cooke, P. and Morgan, K. (1985) 'Flexibility and the new restructuring: locality and industry in the 1980s', Papers in Planning Research, No. 94, Cardiff: Department of Town Planning.

Cooke, P. (1986) 'Modernity, postmodernity and the city', Paper presented at Conference on 'Problems and Strategies for Big Cities: European Perspectives', Copenhagen, 18–21 September.

Citizen, The. (1985) Gloucester, 7 June.

Cowen, H. (1987a) 'The Cheltenham ambience: reading the built environment', Working Paper 2, CURS Cheltenham Study Research Working Papers, Gloucester: School of Environmental Studies, GlosCAT.

Cowen, H. (1987b) 'Defence engineering in the Cheltenham locality', Working Paper 3, CURS Cheltenham Study Reseach Working Papers, Gloucester: School of Environmental Studies, GlosCAT.

Cowen, H. (1987c) *'Farewell to the colonels? Local impacts and economic restructuring in Cheltenham*, Paper presented at Sixth Urban Change and Conflict Conference, International Sociological Association, University of Kent, Canterbury, 20–23 September.

Department of Education and Science (1986) *Higher Education: Meeting the Challenge*, London: HMSO.

Dowty plc (1975–86) *Annual Reports*, Cheltenham: Dowty.

Dowty plc (1986) *Dowty News*, Cheltenham: Dowty.

Fothergill, S. and Gudgin, G. (1982) *Unequal Growth: Urban and Regional Employment Change in the UK*, London: Heinemann.

Giddens, A. (1984) *The Constitution of Society*, Cambridge: Polity Press.

Gloucestershire County Council (1980–86) *Budget and Accounts*, Policy and Resources Committee, December.

Gloucestershire County Council (1984) *Minutes of Meeting, Education Committee, 'Organization of Secondary Education'*, 19 March.

Gloucestershire County Council, Planning Department (1985) *Population Monitoring Report*.

Gloucestershire Echo, Election Results (Cheltenham: 1971, 1976, 1983, 1984, 1985, 1986).

Gloucestershire Echo, August 1979.

Gloucestershire Echo, 17 December 1985.

Gloucestershire Echo, 4 July 1986.

Gloucestershire Echo, 10 April 1987.

Gloucestershire Echo, 1 July 1987.

Gloucestershire Echo, 23 October 1987.

Gloucestershire Echo, Housing Supplement Weekly, 1987.

Hall, P. (1987) 'Flight to the green', *New Society*, 9 January, pp. 9–11.

Hamnett, C. (1987) 'Accumulation, access and inequality: the owner occupied housing market in Britain in the 1970s and 1980s', Paper presented at Sixth Urban Change and Conflict Conference, University of Kent, Canterbury, 20–23 September.

Hart, G. (1981) *A History of Cheltenham* (2nd edition), Gloucester: Alan Sutton Publishing Ltd.

Herzberg, H. (1985) 'Regent Arcade by Dyer Associates: a new promenade for Cheltenham', *The Architects' Journal*, 181, 20, pp. 51–69.

Hurley, J. (1979) 'Capital state and housing conflict: a study of housing in Cheltenham, unpublished Ph.D., University of Warwick.

Independent Schools' Year Book (1986).

Labour Research Department (1986) *Bargaining Report*, London: LRD, July.

Little, B. (1982) *Cheltenham*, London: Batsford.

Little, B. (1967) *Cheltenham in Pictures*, London: David & Charles.

Livingstone, I. (1987a) 'Restructuring and the Cheltenham economy'

Working Paper 1, CURS Cheltenham Locality Study Working Papers, Gloucester: School of Environmental Studies, GlosCAT.

Livingstone, I. (1986b) 'Working for the state: civil servants in Cheltenham', Working Paper 5, CURS Cheltenham Locality Study, GlosCAT.

Lovering, J. (1985) 'Regional intervention defence industries and the structuring of space in Britain: the case of Bristol and South Wales', *Environment and Planning D: Society and Space 3*, pp. 85–107.

McNab, A. (1987) 'Producer services and insurance in Cheltenham in the 1980s'. Working paper 4, CURS Cheltenham Locality Study Working Papers, Gloucester: School of Environmental Studies, GlosCAT

Miliband R., Panitch L. and Saville, J. (eds) (1987) *Socialist Register 1986– 1987*, London: Merlin Press.

Newby, H. (1987) *Country Life: A Social History of Rural England*, London: Weidenfeld and Nicolson.

Pakenham, S. (1971) *Cheltenham: A Biography*, London: Macmillan.

Penycate, J. (1986) 'The property boom widens the North–South divide', *The Listener*, 13 November, pp. 4–5.

Smiths Industries (1977–86) *Annual Reports*.

Shelter Advisory Centre (1986) *Advice Services – The Case for Investment: Report to Cheltenham Borough Council*, Cheltenham.

Thrift, N. (1986) 'Localities in an international economy', Background paper presented at Economic and Social Research Council Workshop on Localities in an International Economy, University of Wales Institute of Science and Technology, Cardiff, 11–12 September.

Walker, F. (1972) *The Bristol Region*, London: Nelson.

West, N. (1986) *GCHQ*, London: Weidenfeld and Nicolson.

Whitaker, J. (1985) *The Best: H. H. Martyn*, Southam, Gloucestershire, Old School House.

Whitaker, R. (1987) 'Neo-Conservatism and the state', in Miliband, R, *et al.* (eds), *Socialist Register 1986*, London: Merlin Press, pp. 1–31.

Wiener, M. (1985) *English Culture and the Decline of the Industrial Spirit, 1850–1980*, Harmondsworth: Penguin.

Williams, R. (1973) *The Country and the City*, London: Chatto and Windus.

Wright, P. (1985) *On Living in an Old Country, The National Past in Contemporary Britain*, London: Verso.

4

Restructuring Lancaster

PAUL BAGGULEY, JANE MARK-LAWSON,
DAN SHAPIRO, JOHN URRY, SYLVIA WALBY,
& ALAN WARDE[1]

Introduction

Lancaster is an area replete with contrasts. From a distance four buildings in particular stand out against the skyline. The Castle, built on a promontory in a bend in the River Lune, is a well-preserved fortification made from dour local stone. The extraordinary Ashton Memorial is an immense folly demonstrating Victorian and Edwardian civic paternalism, designed in a 'reactionary' classical style yet partly built in the 'modern' material of concrete. Bowland Tower at the '1960s New University', is an architecturally cautious and modest tower built of brick, yet dominating the surrounding landscape. Finally, the 1970s/1980s Heysham nuclear complex, with its huge twin blocks located closer to a substantial conurbation than anywhere else in the UK, visually dominates what is otherwise the most beautiful bay in the north of England. These buildings stand as icons of different projects for the Lancaster area: that of a historic centre of administration; that of the dominance of manufacturing employers cowing their workforces into supine acquiescence; that of the public sector optimism of the 1960s; and that of Thatcherism, to continue developing alternative energy sources to coal where considerations of safety and the environment can be seen as lower priorities (although the actual decisions to build at Heysham were taken by Labour).

The area can be symbolized in another way, through a single building. St Leonard's House was originally the Gillows furniture factory, specializing in high-class fittings especially for ocean liners. When the luxury market collapsed and the factory closed in 1962, it provided the first home for the university until it moved to its green-field site (like so much manufacturing industry at the time). And then, since the late 1960s, St Leonard's House has provided the seedbed site for 'Enterprise Lancaster', representing another aspect of

---

[1] Lancaster Regionalism Group, Department of Sociology, University of Lancaster.

the Conservative project where small business is seen as *the* means of economic regeneration.

Interestingly, three kinds of building are missing from the area: large office blocks providing private sector service employment; extensive, council-built, high-rise housing blocks; and private sector built shopping centres. Their absence reflects the relative weakness locally of national and international consumer and producer service companies, and of the local labour movement.

The built environment, because it alters relatively slowly, stands as a marker of uneven development. The buildings trace the changing economic functions of the town and the layers of human experience and social conflict which comprise its history. Lancaster was an established commercial and service centre in the eighteenth century when it was a port for the Atlantic trade, but that trade had declined by 1800. The Lancaster area, on the edge of the Lake District, is tucked into the north-west corner of Lancashire, and was peripheral to its core centres of nineteenth century industrial development in cotton spinning and weaving. Lancaster grew much later, concentrating on the production of the textile-related products of linoleum and oil-cloth. Morecambe, five miles away on the coast (and part of the same local authority district since 1974), developed into a working class holiday centre much later than many of the major resorts like Blackpool and never attracted large numbers of visitors from Lancashire itself. However, by 1920 or so Lancaster had become the major British centre for the production of linoelum, and Morecambe had developed into the quintessential working class holiday resort.

The character of the area in the twentieth century was shaped by a particular structuring of working class experience. In Lancaster semi-skilled un-unionized male workers and an authoritarian paternalism combined to preclude even a working class defensive culture. In 1930 a senior trade union visitor described Lancaster as 'a black hole of trade unionism'. In Morecambe, the petty bourgeoisie serviced the organization of 'rational recreation' centred around the contrived pleasures of the 'family' holiday. This gave a particular social and political fix to the area, indicated for example by the fact that its two constituencies have only once returned a Labour MP. Relatively isolated and rather parochial, the local population was distanced from many elements of the working class culture prevalent in the north-west in the first half of the century.

The deindustrialization of the area developed both earlier and more extensively than elsewhere. It has transformed the locality in some ways for the better, although at the cost of generating new strains and social divisions, most obviously around unemployment and the rundown of Morecambe as a resort. In the past the area had a handful of fairly large manufacturing firms based in two or three industries, which employed mostly male semi-skilled workers, and had considerable business and personal linkages with each other. Their fortunes, along with those of the Yorkshire woollen industry, determined the state of the local economy. Now the local economy is more fragmented, containing education, health, gas field exploration and nuclear power; a number of growing areas of manufacturing employment, generally in small-medium companies in metals and engineering, printing and publishing; and a very depressed tourist industry starved of private investment and also of public investment at a time of local authority budgetary constraint. Previously important manufacturing industries – floor coverings, wall coverings, plastics, artificial fibres, clothing and footwear – now play a much smaller role and in some cases have disappeared entirely.

The character of any particular place depends upon its location in a number of spatial divisions of labour and on the particular way in which those are combined together, in a kind of 'geological' structure with economic, cultural and political components. That structure provides the conditions of subsequent development. The prospects for places like Lancaster are mixed. Though having suffered recently from large-scale national and international processes it is not merely at their mercy: localities possess their own pro-active powers. There is a strong emphasis within a variety of political parties and movements to develop more decentralized forms of politics and organizational structures. As a consequence we should, as Mike Rustin argues, 'expect more of small and medium-sized towns, and not merely compare them disparagingly with the greater cosmopolitan excitements of Manhattan and similar metropolises' (1986, p. 493). This is because

> Most people cannot live in Manhattan or its like, or even visit it very often, and the undue adulation of the marvels of such centres is a symptom of an unduly stratified, mercenary, and centralized cultural market place.... We should favour a culture as well as a politics of decentralisation, and have some confidence that ordinary-sized towns need not be places where one would only live if one has 'no plans of one's own'. (1986, pp. 494–5)

Indeed, recent migration to modest-sized towns, closer to the countryside, often with good public amenities, may well reflect the widespread wish to live in a knowable community, offering a possible vision of a post-materialist, post-modern politics. Medium-sized towns like Lancaster possess, as Rustin emphasizes, an important

Plate 4.1   The Ashton Memorial, Williamson Park, Lancaster.

role in the emerging debate on the nature of a viable urban experience as we move towards the twenty-first century.

Restructuring the local economy

There have been three main phases of industrial activity in Lancaster in the twentieth century, separated by two periods of substantial restructuring (for the Morecambe equivalents see Urry, 1987b). These phases are, first, the beginning of the twentieth century until the end of the 1920s; second, from the 1930s to the mid-1960s; and third, from the 1960s to the present day.

Phase 1: 1905–1929

The first phase is characterized by the overwhelming dominance of two firms manufacturing linoleum and related floor and table coverings: Storeys and Williamsons. In 1911 about 30 per cent of both men and women in Lancaster itself were engaged in the industry; by 1921 the proportion was probably 35 per cent (see Warde, 1988). Lancaster had at the time an extremely dominated labour market, which was also self-contained. Worker quiescence was probably not the product of 'paternalism' as such. Rather, it arose from the importance of the internal labour market which encouraged lifelong obedience. In Storeys, examination of the labour records for this period shows that there was a single port of entry for labourers (below age 18) with very little recruitment of older workers. Since there was very little chance of gaining similar employment elsewhere locally, labour indiscipline would be extremely costly. The two main employers collaborated to control the labour force. One entry in Storeys' leavers' book, giving reason for the dismissal of a certain Vincent Landor in 1907, reads: 'Discharged. We found he had worked for J[ames] W[illiamson] and S[on] and had not left properly (they complained).' They also colluded over wage rates and the prohibition of general labourers' unions (see Warde, 1988).

Both Storeys and Williamsons put resources not into facilities for their workers but into a strategy of civic benevolence, providing voluntary hospitals, educational institutes, public parks and civic buildings. The two benefactors apparently competed with each other to appear the more generous, even contributing to the building of Lancaster Trades Hall! During this period these two employers maintained a local political hegemony (see Mark-Lawson, Savage and Warde, 1985).

Phase 2: 1930–1964

The industrial base began to change in the mid-1920s. A number of firms associated with textile manufacture and processing, and especially the production of the new artificial fibres, were established in the town. The labour market was consequently rather tight with remarkably low unemployment recorded in the 1931 census (6.5 per cent for men). The industrial base had become more varied and the population had expanded during the decade and was to grow even faster than the national average during the next twenty years. There was considerable reorganization of production in the late 1920s. The major form of scientific management introduced into Britain, the Bedaux system, (Littler, 1982), was rapidly embraced by the British linoleum industry. Storeys used Bedaux as consultants between 1931 and 1967, and three other major local firms, outside linoelum production, were also clients. There was thus an important restructuring of Lancaster industry in this period with both a deskilling of labour and the introduction of various schemes of work measurement. By the 1930s Lancaster industry was in the forefront of new systems of work organization. There was scarcely any labour opposition to the introduction of the Bedaux schemes. Absence of resistance was typical of Lancaster. Councillor L. Oakes, President of the National Union of Dyers, Bleachers and Textile Workers, commented after a failed strike at Nelsons Silk Ltd in 1939:

... it is time the people of Lancaster woke up and remedied the conditions, and the only way to do that was to get strong Union representation.

We should not blind our eyes to the fact that some of these people bring their factories to this area, not because they want cheap water, cheap electricity, or cheap gas, but because they want to exploit a tradition of cheap labour – a tradition attached to this City which has got to be swept away.... (*Lancaster Guardian*, 14 July 1939, p. 5)

Nelsons had probably located in Lancaster because of the low wages and its reputation for having a quiescent labour force, and employers sought to preserve a low-wage economy and oppose the unionization of the predominantly unskilled labour force. Some of the firms establishing themselves later in the interwar period, such as Lansils, did introduce welfare facilities but this mainly reinforced the strong internal labour markets found in many of the other firms as well (see Warde, 1988). This prevented labour from taking advantage of periods of tightness in the labour market to bargain up their wage levels. Conditions of employment in Lancaster were the antithesis of the casual labour conditions in the port of Liverpool at the time.

The phase from the 1930s to the 1960s was one of relative prosperity. There was extensive public and private investment, with strong representation in growing industries. Fogarty commented in 1945 that 'the high rate of immigration and population growth before the war is sufficient proof of the prosperity of the area, and there is no reason to suppose that at least equal prosperity will not be resumed after the war' (1945, p. 213).

Indeed, both manufacturing and service employment grew considerably during the 1950s. So did the population, which increased by about 4000 (Murgatroyd, 1981, p. 8), though Lancaster never grew at anything like the rates of Swindon or Kirkby. There was new investment, in chemicals, oil refining, artificial fibres, plastic–coated goods and tourism. Lancaster attracted 7 per cent more new industrial building than one might have expected on the basis of its size and structure (Murgatroyd and Urry, 1985, p. 33). Even as late as 1964, local employers maintained that because there was more than full employment, they were unable to recruit the labour required, and indeed that labour had become much scarcer in recent years (see Fulcher, Rhodes, Taylor, 1966, pp. 40–1). By the late 1950s Lancaster, in marked contrast with much of the rest of Lancashire, had developed an industrial structure highly favourable for future economic growth. As Fothergill and Gudgin note:

'Expressed as a percentage of 1959 employment the worst subregional employment structure for manufacturing was North East Lancashire (−20.0%), reflecting its heavy dependence on the declining cotton industry, and the best [in the whole country] was Lancaster (+19.5%) ... dominated by a handful of firms in growing industries.' (1979, p. 169)

Phase 3: 1965–the present

A massive process of restructuring in the UK in the 1960s transformed the Lancaster area. Lancaster experienced a dramatic 'deindustrialization' of its economy, losing employment in.manufacturing earlier and more quickly than most other areas of the country. But because of its representation in growing industries, this cannot simply be attributed to sectoral decline. Transformations in the organization of manufacturing production under the sway of regional policy undermined the viability of established local firms and industries. Amongst the market leaders in the 1940s and 1950s, they failed to respond to the changed conditions in the 1960s. Simultaneously Lancaster was conspicuously unsuccessful in attracting new manufacturing industry. However, service and other non-manufacturing employment increased. Overall, manufacturing employment fell by 47 per cent between 1961 and 1981 while services employment rose by 28 per cent. Service sector employment in Lancaster grew rapidly in the 1960s, more slowly in the 1970s, but scarcely at all in the early 1980s. Most of this expansion took place in public sector services, although producer services grew in the 1980s. The pattern of service

Table 4.1 *Lancaster: total employment by sector and percentage changes in employment by sector 1971–81 and 1981–84*

	1971 N	%	1981 N	%	Percentage change 1971–81
Primary	1092	2.6	852	1.9	−22
Manufacturing	11227	26.4	9640	21.8	−14
Construction and utilities	3437	8.1	4467	10.1	31
Services	26648	63.0	29258	66.1	9
Total	42604	100.0	44237	100.0	4
	1981 N	%	1984 N	%	Percentage change 1981–84
Primary	746	1.8	790	1.8	6
Manufacturing	9094	21.4	6549	14.8	−28
Construction and utilities	4380	10.3	6770	15.3	55
Services	28328	66.6	30053[1]	68.1	6
Total	42548	100.0	44162	100.0	4

Source: NOMIS, Census of Employment

Notes: [1] The 1984 Census of Employment failed to cover one of the largest public sector service employers in Lancaster and indeed the whole of Lancashire. We have corrected the Service sector data for 1984 using our own information from the employer concerned. Other users of the 1984 Census of Employment should note that the problem applies to Lancashire as a whole (Lancashire County Planning Department, 1988).

growth thus remains very different from that of Swindon or Cheltenham.

The early 1980s have seen very rapid sectoral change (see Table 4.1 and Figure 4.1). Despite the changes in both the boundary of the TTWA and the industrial classification in 1981, the sectoral division of employment in 1981 remains remarkably constant, enabling fairly meaningful comparisons. In the three years 1981–4 the decline in manufacturing employment has been twice as fast as during the previous ten years. In complete contrast, the construction and utilities sector has grown almost twice as fast during the 1981–4 period. This is largely due to the construction of the Heysham nuclear power complex and, to a lesser degree, the development of the Morecambe Bay gas field base. The primary sector of employment resumed growth during the 1980s, albeit from a very low base, and the service sector experienced continued growth in financial services, hotels and catering, retailing and health.

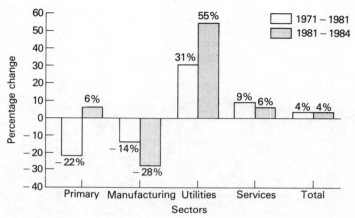

Figure 4.1 Lancaster: percentage changes in employment by sector, 1971–1981, 1981–1984.

The decline of large-scale, manufacturing industry

The decline of the old manufacturing base in Lancaster began with the large-scale redundancies announced at Williamsons in 1967 and culminated with the closure of Storeys' main site in 1982. During that period Williamsons reduced their labour force by 2000 to around its present level (540 employees); Lansils, which had 2000 employees in 1950, cut jobs steadily before closing in 1980; and the medium-sized operations like Nelsons and Standfast Dyers reduced their labour forces too, though they continue to exist. During this period only one substantial manufacturing firm located in Lancaster – Hornsea Pottery in 1972. However, many small manufacturing plants have taken root in the area, the number of firms increasing from 148 in 1980 to 201 by 1986.

The decline of manufacturing employment has had a wide range of effects on Lancaster workers. For many older men it signified the end of work, or of skilled work. An apprenticed engineer made redundant from Lansils in 1981, at age 53, had been unemployed for twelve months and then become a hotel porter: 'because me being an engineer, this type of job is just like you being a doctor and ending up being a bus driver, isn't it'. The whole experience of redundancies was demoralizing, even for workers who did not personally lose their jobs:

'it was incredible, there were some really damn good tradesmen, and yet every month – towards the end we got to saying black Friday, you know – every month you looked on the notice board to see if your name was on

Plate 4.2   Heysham (1) and (2) nuclear power stations.

the notice board. If it wasn't on the notice board, you had your job for another month, and I thought, I'm not having this, so I left.' (former engineer at Williamsons, subsequently a hospital engineer.)

It was not just those who had worked in manufacturing who were concerned. Several of our interviewees considered that Lancaster was a worse place without those jobs; others felt that it changed the nature of the place:

'When we first came and lived around here, when you went out early in the morning you saw the workers going to the Storey's factory and then coming out in the evening, and then suddenly they were gone – it just closed and it seemed a bit like a dream. And so [Lancaster] has changed a lot. I suppose in some senses for the worse and for the better – I'm not one of the workers from Storeys!' (Nurse)

The expansion of the public sector

Lancaster became heavily dependent on public employment, particularly in welfare services (Bagguley, 1986b). The city had been a centre for health care, particularly for mental health, in the nineteenth century. The old County Lunatic Asylum, now a large mental illness hospital, always serviced a large area; and a big hospital for the mentally handicapped was opened in 1868. These facilities expanded after the Second World War, though they are beginning to be run down now in line with policies of 'community care'. Successive NHS reorganizations have led to the expansion of services at the general hospital. By 1981 the health services had become the largest employer in the area and by 1985 the NHS employed 5442 people. The education sector also expanded. The university was founded in 1964, the year after a new college of education. By providing for people outside the area this sector too came to be proportionately large, and by 1984 it employed 4956 people.

In 1967 it was announced that one of the first AGR nuclear power stations was to be built at Heysham. Its construction in the 1970s, and the building of a second reactor on the same site in the early 1980s, provided badly needed employment in the Morecambe area. The second station was the largest construction project in the UK in the early 1980s, employing over 6000 workers at its height. Public sector employment has been vital to the area since private sector investment has been limited. Indeed, in a whole range of ways the state has been deeply involved in the restructuring of Lancaster (Shapiro, 1988b).

Explaining industrial restructuring

Although from the point of view of Lancaster the effects of restructuring have been contingent, there are powerful underlying mechanisms which do entail general changes in the spatial division of labour. There is a relatively constrained set of processes by which industrial sectors come to be restructured. Ideally we should be able to explain why different patterns are to be found in different sectors. The current patterns of restructuring have been shaped by five key determinants: internationalization, government policy, changed gender relations, new managerial strategies and shortage of land.

INTERNATIONALIZATION

The UK economy, as both a leading exporter and a leading recipient of capital, occupies an important site in international economic processes. Between 1971–83 the real value of foreign direct assets rose by nearly 45 per cent with manufacturing and especially services showing marked increases. Simultaneously UK manufacturing firms reduced investment and employment in Britain. Manufacturing firms historically based in the north have shown the greatest propensity to shift the balance of production overseas. Two key processes of internationalization have been relevant to recent industrial restructuring in Lancaster. Firstly, *import penetration* has severely affected clothing and footwear in Lancaster, encouraging them to go 'up market'. The second feature of internationalization – the place of Lancaster production units in the *global strategies* of the multinational companies that own them – has been more important. In the later 1970s four UK based multinationals – Courtaulds, ICI, Turner and Newall, and Unilever – controlled approximately two thirds of manufacturing employment in Lancaster. By the mid 1980s three of these (ICI, Turner and Newall, and Unilever) no longer owned any production sites in Lancaster whilst Courtaulds maintained only a modest presence. These companies have been amongst those centrally involved in reducing their UK workforces and shifting production overseas. Between 1975 and 1982 UK employment in these firms declined by 43 per cent in ICI; 49 per cent in Courtaulds; 20 per cent in Unilever and 21 per cent in Turner and Newall (Lloyd and Shutt, 1985).

The largest manufacturing firms in the Lancaster area were most affected by two processes of restructuring: *investment and technical change* and *rationalization* (Murgatroyd and Urry, 1985). ICI's Heysham site was peripheral to their needs. Investment and technical change especially at the ICI Billingham works on Teesside in the 1960s and early 1970s meant that the Lancaster plant was too small and technologically backward to compete and it closed in 1986. Rationalization characterized the linoleum industry due to the growth of new products (cheap carpets made from artificial fibres and plastic based floor coverings). Of the two main companies, Storeys, in the 1950s managed to switch to PVC plastic-coated products and as a result sold off their cotton spinning and weaving mills which had provided the backing for linoelum. Turner and Newall bought Storeys of Lancaster in the late 1970s and, in a fit of rapid rationalization, disposed of its production sites to another Lancashire firm, which promptly made most of the employees redundant. Williamsons by contrast had continued with linoleum and when the market collapsed in the early 1960s they were forced into a merger with Nairns of Kirkcaldy. Major redundancies followed in 1967 and they later switched out of linoleum altogether. Unilever took over the Nairn group in the 1970s and employment has slowly declined until Nairn's was sold to a Swiss-based multinational in 1985. The site for the moment appears secure. Courtaulds closed its Lansils complex in Lancaster in 1981, centralizing its UK production of artificial fibres in Derby. In the case of Turner and Newall and Unilever, Lancaster sites have been peripheral to the companies' key production sectors. With ICI and Courtaulds there has been a centralization of production elsewhere in the UK at the expense of Lancaster.

The restructuring of the large manufacturing plants was part of the global strategy of the large corporations. Their decline locally had a knock-on effect on dependent producers of intermediate goods in the area, for example in chemicals. New demands for intermediate products have, however, developed. Engineering firms expanded to meet the requirements of the CEGB and their contractors completing the building of Heysham 1 and 2. By 1984 there were over 60 engineering supply, design and general engineering companies in the Lancaster area. There were also some newly established companies, particularly in the manufacture of paper and board, printing and publishing. There seem to be few clear reasons given for locating in Lancaster. For example, a local high technology company which designs and produces wafer testing equipment for the semi-conductor industry moved to Lancaster in 1984 from Macclesfield (and then took over a company based in Milton Keynes in 1986). It saw the main advantages of Lancaster as being modest rents, good cheap local subcontractors, and proximity to the motorway network (halfway between the 'M4 corridor' and 'Silicon Glen' in Scotland!)

THE EFFECTS OF CONSERVATIVE GOVERNMENT, 1979–88

Government policy has had four major consequences on the restructuring of the Lancaster economy:

1. Monetarist policies nationally, according to Martin, caused at least half the job losses between 1979 and the early 1980s (1986, pp. 258–64). Nationally in manufacturing, this was reflected in a 50 per cent increase in excess capacity between 1979 and 1982, a reduction in the output index from 109.4 to 94.5, a doubling in the number of confirmed redundancies, two and a half times as many company liquidations, a reduction by a third in the level of capital investment, and an increase in regional differences, with the north-west suffering one of the largest reductions in manufacturing output and employment between 1979 and 1983.

2. The size and geographical scope of regional policy has contracted. There has been a steady reduction in the value of regional aid since 1978/9; in real terms, it fell from £842m to £560m in 1983/4 (Martin, 1986, p. 272). Lancaster only had Intermediate Area status and so was not eligible for the more important Regional Development Grants. With the restructuring of regional policy in 1984 Lancaster lost Intermediate Area status (it had attracted Regional Selective Assistance worth about £2.5 million since the early 1970s; Urry, 1987a, pp. 5–6). The shift in government policy away from 'spatial' to 'sectoral' support (in 1982/3 worth three times the value of regional policy) has seriously weakened those places without the relevant sectors.

3. The powers of the trade union movement in general, and especially those of the older craft unions, had only really developed since the 1960s – by the 1980s their influence had disappeared. New manufacturing firms established in the 1980s were mainly un-unionized – while the strength of the local union movement was to be found in education, health and nuclear construction.

4. The state has been restructured with a proportionate shift away

from education and health expenditure and an increase in law and order and defence expenditure. Hence employment in Lancaster in the first two areas has not continued to expand. Other localities, like Cheltenham, which are more dependent on defence expenditure and private education, have correspondingly expanded employment.

Restructuring in the health services, for example, has occurred locally partly as a consequence of the government's policies of privatization. A private hospital, part of the Nuffield Hospital foundation, has opened, its main effect being that some staff have moved from the public sector. But the most significant development in private welfare services in Lancaster has been the provision of residential homes for the elderly. Overall these developments have been stimulated by changes in DHSS regulations in the early 1980s, but in Lancaster the growth has been significant. In December 1983 there were 29 private residential homes for the elderly in Lancaster District, providing 540 places, and 18 of these had opened during 1983. By the end of 1986 there were 41 such homes with 655 places. At the same time public sector provision remained static. This may partly reflect the high proportion of elderly people living in the Lancaster district (Bagguley and Shapiro, 1986); the declining tourist demand for hotels in Morecambe opening up suitable buildings for conversion; and the presence of a trained workforce from the substantial local public health service.

Another feature of privatization is the competitive tendering for ancillary services. The local health authority has had to be pragmatic in implementing this programme in Lancaster largely because of the lack of competitive tenders from the private sector. Tendering for the major services was not completed until late in 1986, and most of the tenders went 'in-house' since the major private sector firms which have successfully tendered for contracts elsewhere in the county concentrate their operations in the large conurbations. The main impact of tendering on the staff in Lancaster has been to intensify the work process. A hospital cleaner described the deleterious effects on the quality of her job of a successful in-house tender:

> 'We've been knocked down to twenty-five [hours per week] ... it's terrible, it's bad, you do a hell of a lot more in twenty-five hours, oh yes, it's gone down, definitely gone down, I mean there is supposed to be two to a ward, all right there is, but when you have had your days off, like I've had four off now, my mate will be on her own, and its hard.'

Whereas there had previously been relief domestics who covered for absentees, they had been dispensed with. The consequences for the cleaners is brought home in the account of a day's work, paid at the rate of £1.90 per hour.

'I go in at eight o'clock, wash all the pots, early morning drink, pots, cups and saucers, there's sixteen grown-ups on our ward everyday here, it's a skin ward, and they are all good eaters, because they are not in because they are ill, it's because of their skin, and we do all the pots and then we have to do the day room and the smoke room, and then hoover the hallway, come back into the kitchen, for the odd pots that have come in after time, get the trolley fixed, put that outside for the porter to take, do the washbasins in the ward, empty all the rubbish, and that brings you up to ten o'clock, then we have a break, ten minutes, then we come back, all the coffee pots are there then, and we do all the coffee pots again, set the trays up for the dinner, then my friend goes into one end, I go into the men's end and she goes into the ladies' end, she does the ladies' toilets, I do the men's toilets, do all the ward, sidewards, sister's office, X-ray departments things like that and when you look at the clock it's ten to twelve. Dinner is at twelve then when you come back it's murder, you've to start then and it's like a full hour washing up and all the kitchen, mopping over the floor and everything, leaving everything clean and that brings you to half past four and you sometimes crawl out on your knees, because as I say if your mate is off you've to do it all on your own, the whole lot, it's terrible. I think Maggie Thatcher should get herself up here and have a look, in fact I would give her my end to do one morning, just to let her see what it is all about.'

CHANGING GENDER ENVIRONMENT

Probably the most important change in the social division of labour in Britain in the last 25 years has been the increased involvement of women in paid employment. In the early decades of the twentieth century, though there were striking local variations, few married women were economically active. Since the Second World War local variation has evened out and married women have become a major fraction of the labour force. This reflects a range of social and economic developments: greater demand for female labour, an expansion of jobs traditionally filled by women, greater educational opportunity, smaller family size, women's movement struggles, equal opportunities programmes, etc. Now, the availability of reserves of female labour is a key feature of any local labour market and economic restructuring cannot be understood without close examination of gender divisions in employment practices.

In Lancaster, as in other localities, women do different jobs to men. The very fact of occupational segregation makes it difficult to disentagle those changes in women's employment which result from processes specific to an industrial sector (e.g. the absolute decline of employment in textile manufacture, where many women worked) from those that are a consequence of patriarchal intervention (e.g. a trade union insisting upon men's jobs being preserved at the expense of women). Both types of process have occurred in Lancaster. One interviewee, a manager, recalling employment change in the lino-leum industry in the 1950s, illustrates the latter. As new processes were introduced men were transferred from weaving operations to the new processes for producing plastic coated products, while women were declared redundant:

> *Interviewer*: Were the women who were weavers made redundant?
> *Respondent*: Yes, certainly yes . . . equal opportunities didn't enter into it . . . manufacturing PVC is quite physical. I don't doubt that you could find a Russian-built type that would laugh at it . . . the emphasis on women's employment in [this company] was with the inspection, wrapping, . cutting up, packaging . . .
> *Interviewer*: And the Unions didn't bother about that?
> *Respondent*: . . . Well, they weren't bothered and at one stage, I mean, we had one lady who was a shop steward, so well ahead of the equal opportunities we had a female shop steward, no there was never to my knowledge, there was never any pressure . . . about women not being employed in the manufacturing processes. And also it was full-time shift work, twenty-four hour, seven days a week which creates a deal of problems about it for employing women.

Patterns of segregation by industry account for some of the differential effect of economic restructuring on the gender com-position of the labour force. For instance, in the health services in Lancaster the proportion of employees who were women increased from 70 to 74 per cent between 1971 and 1981 (Bagguley, 1986b, 26). However, the position of women deteriorated in the sense that the higher levels of nurse management have been taken over by men, partly because the Equal Opportunities legislation of the early 1970s opened up an occupational group previously for 'women only'.

During the first two phases of Lancaster's industrial experience the female economic activity rate was low compared with both the UK and Lancashire. As elsewhere, the movement of women into the local workforce which took place during the Second World War had little lasting impact. After the war economic activity rates for women, and men, remained low. However, between 1951 and 1971, while the male economic activity rate increased slightly from 81 to 85 per cent, the female activity rate rose dramatically from 32 to 53 per cent. However this does *not* indicate 'feminization' in the sense of women moving into forms of employment previously dominated by males.

Two main tendencies linked to the rise in the female activity rate are, first, the considerable growth in part-time employment and, secondly, the increase in the married women's activity rate. Between 1951 and 1981 female employment rose by 5400, some 40 per cent, and in the years 1971–81 the women's economic activity rate increased by 18 per cent. The most dramatic increase in this latter period has been in the married women's activity rate, although there has also been an increase for single, widowed and divorced women. The 1970s also saw considerable growth in part-time work, with a percentage increase for women of 39.5 per cent between 1971 and 1976 (Bagguley, 1986a).

However, to describe this as 'feminization' is misleading. In manufacturing, there has been little substitution of female for male workers; if anything a process of *defeminization* has taken place as female labour has been shed. The increase in employment opportunities for women is accounted for almost entirely by the public service sector, and here there has been no 'substitution' of male workers by female workers. Instead, traditionally female occupations have expanded, particularly in part-time work: public consumer services increased its proportion of total service employment in Lancaster from 40 to 43 per cent between 1971 and 1981 and the proportion of women employed in this sector increased from 57 to 62 per cent.

In Lancaster women have always been well represented as nurses and teachers, but while women's 'opportunity' to enter low paid, low level and insecure occupations has opened up since 1971, their position within the management hierarchies has worsened. This is a national phenomenon with women's managerial positions within traditionally female professional areas challenged by the Sex Discrimination Act. In Lancaster, where the health sector is the major employer, the existence of a large mental health and mental handicap sector means there are more male nurses seeking promotion than there might be in other areas. On the major site for acute care, for instance, of four key nursing management positions, three are now held by men. 'Masculization', combined with an expansion of the numbers of traditional female jobs, is a better summary of these processes.

CORPORATE STRATEGIES: FLEXIBILITY IN THE 1980s

Developments in the strategies used by management to control labour are a further basis of economic restructuring. Recent debates have shifted to a consideration of labour flexibility or the flexible specialization of firms. Two general types of labour flexibility have been identified: (1) functional flexibility where workers move between jobs; and (2) numerical flexibility where workers are employed part-time or on temporary contracts so that employers can match labour input to demand (NEDO, 1986). Much of the previous analysis of flexibility has concentrated on manufacturing industry (e.g. Piore and Sabel, 1984; Tolliday and Zeitlin, 1986). In Lancaster we have found flexibility to be more significant in services than in manufacturing and we believe this may also be the case elsewhere. In Lancaster's manufacturing sector we found only two firms pursuing flexibility with any degree of determination. In one of these instances the firm had cut its semi-skilled workforce to the bone; in the other, most operatives were semi-skilled, and had been functionally flexible for some time, but craft flexibility was being actively sought. Elsewhere in manufacturing nothing like the managerial strategies and organization structures of the 'flexible firm' identified by NEDO (1986) were found.

By contrast, in the service sector, numerical flexibility in the form of part-time employment is central to contemporary restructuring. In Lancaster's hotel and catering sector for example, women's part-time employment doubled between 1971 and 1981 while women's full-time employment fell by almost half. In private consumer services more generally, women's part-time employment in Lancaster increased by 1000 between 1971 and 1981 while total employment *fell* slightly. Temporary employment has always been important in Morecambe's tourist industry, but more recently temporary contract workers for teaching and research have increased as a proportion of higher education employees. Functional flexibility is also important in services. In the large local retail stores staff are trained to operate in a variety of jobs, from working at tills to specialist serving and preparation. In hotels and catering functional flexibility amongst kitchen, cleaning and service staff has developed strongly as a major labour management strategy since the early 1960s (Bagguley 1986a, 1987).

Our research in Lancaster suggests the focus of the flexibility debate on manufacturing has been somewhat misplaced. More innovations have been emerging in services where opportunities for flexibility are more widespread. We have also found a clear division by gender in relation to flexibility. Men are more frequently found in functionally flexible situations while women are predominantly found in numerically flexible situations. This emergent division between functionally flexible workers at the core and numerically flexible workers on the periphery may be giving rise to new forms of labour market rigidity. Ironically it is exactly this rigidity that labour flexibility and current government policies have unsuccessfully attempted to overcome.

THE AVAILABILITY OF LAND

Some argue that shortage of land and hence of available sites in cities has been the major cause of the urban–rural shift in industrial location (Fothergill, Gudgin, Kitson, Monk, 1986). However, in the past this was not a problem for Lancaster. Although in the case of one of the major companies, Storeys, lack of land in their central city site caused considerable operating difficulties in the 1960s and 1970s, it was able to expand on another local site. The local authority has been diligent in the provision of land and accommodation. The city council made great efforts to ensure suitable sites for the university in the early 1960s and for Hornsea Pottery in the early 1970s (in the case of the former in competition with Blackpool where the land offered now houses the zoo!). Since Lancaster is a small free-standing city it has been relatively easy to find sites for development on the edges of the built-up area.

Labour market and social structure

The processes described thus far imply major changes in the kinds of jobs available in the local economy. The labour market is relatively isolated and self-contained with only 7.9 per cent of those living in the district (which more or less coincides with the TTWA) employed outside it in 1981, and 6.8 per cent of those employed inside it living outside.

Overall, the most significant changes in the local labour market between 1971 and 1981 were the increased participation of married women (up 10 per cent) and the growth of part-time women workers (up 8 per cent). Part-time work expanded faster in the first half of the decade (1971–6) than in the years since, and there is no evidence of substitution of part-time for full-time participation, since the latter increased marginally too (Shapiro, 1988a).

Unemployment and 'over-employment'

Although Lancaster went into the recession later than other parts of the country (1982/3 rather than 1979/80), registered rates of unemployment have risen sharply. Redundancies declared between 1981–5 stood at 4500, mostly in manufacturing. In 1981 the official unemployment rate for the TTWA was close to the national average and lower than that for the county. Since then it has moved higher than both. By March 1987, Lancaster was regularly experiencing a higher percentage increase on a high unemployment base. These rates are, though, highly variable within the TTWA with central parts of Morecambe and Lancaster experiencing particularly heavy long-term and youth unemployment.

There is, then – as elsewhere – a 'split community' with in-migrants in secure, well-paid jobs living in the rural areas, while local working class people in the central urban areas are unemployed. Although we have noted the increase in women's employment

147

relative to men in the 1980s, there has also been a striking increase in female:male registered unemployed. Between 1981 and 1985 there was an increase in male unemployment in Lancaster of 28 per cent, and for women of 71 per cent compared to national increases of 27 and 44 per cent. It is possible, using the OPCS 1 per cent Longitudinal Study (LS) to gauge the effect of declining sectors of unemployment. For example, male employment in the 'chemicals' sector in Lancaster declined by 69 per cent between 1971 and 1981. Of those 1971 male chemical workers who were under 65 in 1981, about 85 per cent were still in Lancaster. Fully two-thirds were either unemployed or permanently sick. All of the younger age-group (aged 16–30 in 1971) were still in Lancaster and were unemployed. Of the older age-group (aged 31–54 in 1971) about two-thirds were in employment, dividing roughly evenly between those who had moved up and those who had moved down from their 'firm-specific skills'.

Yet, skill shortages were reported by several firms, for example in: (1) skilled manual work, especially in construction and electrical and maintenance engineering due to competing demand from the nuclear construction industry; (2) skilled clerical/secretarial work; and (3) skilled 'female' work in clothing. Historically the first and second types of occupation have been under-represented in Lancaster. Some firms have recently begun new apprenticeship schemes, suggesting a genuine shortage of skilled labour. The personnel manager for one manufacturing firm said:

'We have a current problem that we can't satisfy and that is, we want to expand the tool room but we just can't get the type of tool maker that we want for love nor money. We advertise nationwide, Scotland, Wales, virtually all the British Isles, except Ireland, we just can't get them for love nor money. We have plenty of applications but they are just so special, and so accepting the situation as it is now, the company has decided to go for more apprentices so that over a long period we will eventually expand the tool room.'

On the other hand, the university, an employer paying at the higher end of local wage rates, reported no difficulty in filling jobs except some very specialist, technical ones. Certainly, as the economy has diversified in recent years lack of sufficient pools of experienced labour may have become an even greater problem than before. This was even so in the case of hotel and catering, according to the manager of the local four-star hotel.

Occupational structure and class formation

Lancaster has a 'polarized' occupational structure with professional/ managerial and semi-skilled and unskilled manual classes over-represented (see Table 4.1). This has been true throughout the twentieth century. This polarization is reducing for men, and it is the distribution of female employment that now makes for the distinct-iveness of the occupational class structure in the Lancaster District. There are considerably more women in professional and managerial classes (I and II) than nationally, reflecting the local importance of the health and education sectors. There are fewer women locally than nationally in IIIN (routine non-manual), reflecting (a) the fact that there are no major private companies with their headquarters based locally; (b) the relatively low proportion of clerical labour in public sector services; and (c) the insignificant amount of employment in 'producer services'.

Analysis by 'socio-economic group' (SEG) in 1981 shows that Lancaster District was *under*-represented in the proportions of employers and managers in large establishments, self-employed professional workers, foremen, skilled and semi-skilled manual workers (see Bagguley and Shapiro, 1986, p. 57). It was *over*-represented in the proportions of employers and managers in small establishments, non-self employed professionals, ancillary workers and artists, personal service workers, unskilled manual workers, own account non-professional workers and farmers. Own account or the self employed made up 13.6 per cent of the employed population in 1981; and it is certain that this figure is now considerably higher. Over the period 1961–81 the main increases for men have been professional employees (increased locally 63 per cent faster than nationally) and unskilled workers (increased locally by 60 per cent while it decreased nationally by 33 per cent).

Occupational structure is one of the principal bases of social class formation. The principal classes in Lancaster are a professional and managerial class predominantly in the public sector and including an unusually high proportion of women, a relatively unskilled manual working class, and a substantial petty bourgeoisie. The experiences of these classes have been very different. The manual workers and the lower echelons of the professional class are more likely to have lived in Lancaster through its changes. For many of the working class this has been a hard period. One household in which the husband, now 53, had been made redundant as an engineer and was working full-time as a hospital porter, and the wife was a part-time nurse declared: 'We just get through, just about. And we had to either give up going out or give up the car, so we give up going out. . . . I don't

Table 4.2 Social class of usually resident economically active population in private households

1981
10% SAMPLE

% of economically active population in social class

	I	II	IIIN	IIIM	IV	V	X	TOTAL
GB all	3.7	20.9	21.9	24.3	18.0	6.4	4.8	100.0
Lancashire all	3.1	20.9	21.0	25.4	19.6	6.3	3.6	99.9
Lancaster District all	4.0	23.6	19.0	23.2	19.2	7.4	3.7	100.1
LD-GB quotient	1.08	1.13	0.87	0.95	1.07	1.16	0.77	1.0
GB % men	89.5	62.5	31.3	87.1	54.7	58.6	64.8	61.1
Lancashire % men	91.0	60.0	31.2	84.8	52.4	62.5	57.4	59.8
Lancaster District % men	86.4	55.6	32.7	84.7	54.6	58.1	63.9	59.5
LD-GB quotient	0.97	0.89	1.04	0.97	1.00	0.99	0.99	0.97

Source: Census of Population

think we are living good really'. The professionals, particularly where there are two earners in the household, are more likely to be content with their material conditions and to judge the quality of life locally in terms of the pleasant environment and improved consumption facilities.

Geographical mobility and political mobilization

The OPCS 1 per cent Longitudinal Study shows that of men resident in Lancaster, economically active, and under 55 in 1971, 71 per cent were still in Lancaster in 1981. There was little variation by social class, and the south, south-east and south-west regions were the destination for only 3 per cent of the movers. In the case of women, the mobility of those economically active and under 50 in 1971 was less than that of men, but again there was no significant differentiation by social class.

The picture for inward mobility – those who were resident in Lancaster, economically active and below retirement age in 1981 – is quite similar, with 70 per cent of men and 73 per cent of women having been resident in Lancaster in 1971. More class differentiation is visible: for men, only 44 per cent of professionals and 61 per cent of intermediate workers were in Lancaster in 1971, compared with 79 per cent of skilled manual and 77 per cent of unskilled workers. For women, 60 per cent of professionals but 78 per cent of semi-skilled workers had been in Lancaster in 1971.

These patterns reflect the fact that the markets for different kinds of labour clearly have different spatial scales and for many categories of the workforce there is no 'local' labour market at all. In Lancaster there are three major sectors – health, education, and nuclear construction and operation – where significant categories of employee are in a regional or national labour market. However just because there is such a market there is not necessarily a great deal of geographical mobility: university teachers are a case in point. Nor is it necessarily those in 'middle class' occupations who are geographically mobile. In Lancaster one of the *most* mobile groups has been manual workers employed in the construction of Heysham 1 and 2. About 70 per cent are manual workers and over half came from outside the TTWA.

Savage (1987) has identified an overall decline in the rates of geographical mobility by about 2–2.5 per cent per annum between 1971 and 1982. He suggests that where there is a well-developed professional/managerial internal labour market, men, once resident in an area, are unlikely to move outside that area (often regional, and defined in house price terms). Such people are likely to develop more of an attachment to the locality/region than previously: being less 'cosmopolitan' in orientation they are much *more* likely to engage in locally based organizations. This pattern may be even more pronounced in areas like Lancaster where movement to the south is, for house-price reasons, particularly difficult. This may partly explain why professional/managerial groups appear to be so strongly committed to the Lancaster locality and local politics: there is what one might term the 'localization of the service class'.

It remains much disputed whether the rate of geographical mobility affects capacities for class action. Some claim that long-term residence encourages class consciousness, others that the geographically mobile will be more politically active and aware. We were not in a position to resolve such claims, though there is evidence that the conservative politics of both Morecambe and Lancaster was diluted, if not totally diverted, by in-migrants. The workers at the Heysham Power Stations have revived Labour Party politics in Morecambe. One CEGB maintenance craftsman, who had moved into the area in 1982, explained his new involvement in Labour politics:

'Back where I lived ... we always had a Labour MP ... and they consistently elected Labour councillors and it was so strong you would never even think of joining the Labour Party, you see. It was only when you come up here and realize how bad it is that you decide that you want it to be more like it was before.'

Local politics

Parties and places

Employers were dominant and workers quiescent in the first two or three decades of this century. This might have changed with the round of industrial restructuring in the 1920s, which saw several new employers arrive in Lancaster. Labour was, however, so weakly organized that it could not mount an effective challenge to the introduction of new forms of scientific management at work or to persistent employer control over the local labour market. It was only in the late 1930s that the Labour Party even began to contest local elections on any scale, and then its successes came through mobilization around consumption issues – especially housing – rather than around industrial grievances. Consequently Labour was not in a position to make an impact at another key point, the 1945 election, which was won comfortably by the Conservatives. Throughout the 1950s and 1960s the Lancaster constituency remained one of the most glaring anomalies to the British pattern of class–party alignment. In 1955 the Conservatives took 11 per cent more of the vote than would have been predicted from the class composition of the city, making Lancaster the most deviant, pro–Conservative constituency in Britain (Piepe *et al.*, 1969).

This began to change during the 1960s as a complex outcome of industrial restructuring which disrupted the local hegemony of the Conservative Party. It is ironic that deindustrialzation was the context in which the Labour Party began to improve its performance in local and national elections. By the late 1960s the political allegiance of workers in the major manufacturing establishments was becoming more varied. The general unions had finally managed to establish themselves in the larger companies by the early 1960s, and a sample survey conducted in 1967 showed no excessive 'Conservative' sentiments among the workers being made, redundant at Williamsons (Martin and Fryer, 1973). The system of internal labour markets had also broken down with quite high rates of movement of manual workers between the main manufacturing plants. The impact of the Labour Party was most sharply registered in the 'landslide' in the 1971 local elections. However, this was less to do with workplace conditions than with the mobilization, again, of discontent in the sphere of consumption – housing and education. The tenor of Labour politics owed more to the involvement of public sector workers in the newer service industries than to the manufacturing working class. Public sector unions were closely involved in the reinvigoration of the Trades Council; many of the Labour activists, candidates and councillors were employed in the public sector; and many of the overlaps in personnel between the newer social movements and the Labour Party were people employed by the state, particularly in health and education.

The Conservative Party's domination of the council was severely threatened in the early 1970s, but its decline was interrupted by local government reorganization which amalgamated Lancaster with Morecambe and Heysham MB. Morecambe was a bastion of the small-employer/self-employed type of Conservatism, but amalgamation also transformed the dimensions of political alignment. A strong Lancaster versus Morecambe split developed in local politics. Issues became 'territorial' rather than 'partisan'. Lancaster councillors argued for spending on jobs and industry, while Morecambe councillors wanted infrastructural spending on tourist-related projects in Morecambe and vehemently opposed plans to develop Lancaster's tourist potential.

The district election results in 1987, in which Labour gained 6 seats and the Alliance 3, went against the national trend and moved majority control from the Conservatives for the first time. Undoubtedly the central issue in this campaign was a proposed Lancaster Plan, and especially its recommendations for a substantial expansion in retail space in the city centre. This directly threatened commercial interests in Morecambe and led to the formation of a new party, the Morecambe Bay Independents. The Lancaster Plan Action Committee brought together many conservationists, market traders and small shopkeepers. The Labour Party also came out strongly against the Plan. This may well be related to the lower levels of mobility for many middle class professionals: in the 1960s such people would have been thought of as 'spiralists' and having relatively little interest in 'preserving the local' (see Bell, 1968).

The 1987 local election results demonstrate the continuing shift against the Conservatives in Lancaster City itself. The Conservatives have steadily lost seats and votes, falling from 45 per cent and 10 seats in 1973, to 25 per cent and 4 seats in 1987. The Labour vote has remained at around 50% but its number of seats has risen from 11 to 14. Other parties, especially the Alliance, have increased from a tiny 1 per cent to 26 per cent over this same period.

Social movements

The political culture of the city has been markedly changed by the flourishing of a variety of social movements, partly outside mainstream party politics. One was opposition to nuclear power. There had been surprisingly little opposition in the 1960s given that the proposed power station at Heysham was to be sited closer to a conurbation than any previous station had been (and any subsequent one in the UK), but opposition did develop in 1971/72. A local group, Halflife, was set up by a university lecturer who had been involved with such groups in the USA, and a number of other Halflife groups followed the Lancaster example. Halflife itself fed into the setting up of Friends of the Earth nationally. In Lancaster many of those involved are drawn from the 'education' sector, especially the university, with its national, and in part international, orientation. Indeed radical politics in Lancaster was in part 'imported', especially from American campuses in the later 1960s and early 1970s (when the Lancaster campus was regarded as one of the most radical in Britain).

Second, Lancaster became well-known for the strength of its women's movement. Before the 1970s there had been little political organization of women around, for instance, welfare issues, compared with other Lancashire towns (see Mark-Lawson, Savage, Warde, 1985). From about 1970 there has been a marked increase in such activity. Radical feminism in the 1970s and 1980s produced a

Women's Refuge, a Rape Crisis Line, an Incest Survivors Group, a Women's Centre, a Lesbian Line, a Lancaster Women's Newsletter, and various cultural events such as the Lancaster Feminist Book Fair, Lancaster Women's Conferences, writing groups etc. Socialist feminists, working through the channels of the local labour movement, have not been so strong. There was a successful campaign around improved maternity provision in the early 1970s and some involvement in workplace disputes. On the whole, though, both socialist and liberal feminism, which might have been expected to produce campaigns around issues of equal opportunities in the workplace and around levels of public service provision – campaigns present in other localities like Liverpool and Thanet – have not been much in evidence in Lancaster.

The Lancaster feminist movement has been strongly 'separatist' (or 'mutualist'), making few formal demands on the local state and political institutions. There are two reasons for this. First, the various local authorities have been unresponsive compared with the metropolitan counties and some London boroughs. Where demands have been made, for instance for funding the women's refuge and for assistance in rehousing the women there, the authorities have been indifferent or hostile. This 'impermeability' may well have reinforced 'separatist' gender politics. Second, much radical feminist politics is linked to the university in various ways. As with the anti-nuclear movement many of these individuals will look to a trans-local national or international community. Again there is evidence that the lack of mobility may enhance the involvement of people in such local political issues (Armour, Mark-Lawson and Walby, 1988).

Local policy

It is paradoxical that although the Lancaster area has been consistently Conservative at both local and national elections, considerable innovation occurred in local economic planning and policy in the 1960s and 1970s (see Urry, 1987a). An active interventionist policy was pursued. This did not come from the commitment of major employers or the political parties to a 'pro-growth' policy as in, say, Swindon or Middlesbrough. Rather the historic weakness of the labour movement locally, and the despair amongst employers at early signs of 'deindustrialization', meant that a powerful grouping of permanent officials was able to carry through a policy of intervention much earlier than in most other places in the UK.

In the early 1960s there were two main developments. First, the university was attracted, partly due to the ability of the Town Clerk

Plate 4.3 Bowland Tower and Lancaster University.

to find a green-field site of 200 acres almost overnight (and incidentally to redraw the city boundaries; McClintock, 1974). Second, there was the designation of the first industrial park in the area, also in 1964. However, by 1966/7 there was real concern for the state of manufacturing in the area and an action group was formed to fight for government help. In June 1967 a more specific plan was announced involving the university, the chamber of commerce, the Town Clerk and various other officers. The idea was to establish an agency, Enterprise Lancaster, which would provide sites for new, mainly science-based firms, often having links with the university. Although there was little Party controversy over this plan, there was recognition locally that the regional policy proposals then being introduced by the Labour government would weaken those places like Lancaster that were ineligible for grants, while Labour-voting industrial cities and regions would apparently benefit. As we have seen, the restructuring of the 1960s and early 1970s, inspired by the regional policy, did indeed cause severe problems for Lancaster.

Enterprise Lancaster was established in 1967 and the first Industrial Coordinator noted that since this was an almost unique development *at the time*, it 'created a lot of interest . . . we had people coming from all over the country, and abroad' (interview, D. Kelsall). there were no models to follow, but the thrust of the initiative gradually came to be provision of seedbed premises in what had been Gillows furniture factory. Good quality initial accommodation was provided at fairly low rents to embryonic firms. The building was conveniently located and there was some provision of common services. The plan envisaged new companies in the research and development, design, and high technology fields, an expectation only partly fulfilled. Some of the firms starting there did grow, though, and Enterprise Lancaster in time also organized transplantation to other industrial estates. By 1984, the companies based in this seedbed and those that had moved to other estates employed 250–300 people.

Industrial estates had been developed by finance under the 2p rate, from the EEC under the Textile Programme, and from central government under the Derelict Land Act. Enterprise Lancaster is also concerned with organizing the movement of tenants and purchasers between various industrial estates; with organizing a rent grant scheme for new starter companies; with direct funding of enterprises through the provision of small-scale industrial loans and mortages; and with general promotional work for the area. It emphasizes the pleasant environment especially towards the Yorkshire Dales and the Lake District; the current high level of offshore and nuclear investment (£4000 million); the non-militant workforce (0.35 of the UK average for days lost through strikes); and location close to both the motorway network and the main Euston–Glasgow line. The disadvantages which the promotional literature tries to offset are the lack of any regional assistance since 1984, the low proportion of skilled workers in the labour force (16 per cent lower than the county rate), and the geographical and social distance of the area from the main centres of economic and political decision making. In general the main thrust has been to concentrate on helping existing firms to grow and bringing about diversification. As the current Industrial Coordinator says, there are 'probably more long term benefits through helping local companies set up and getting started' than in trying to attract large outside companies to invest.

There are other local agencies. *Business for Lancaster*, set up under the auspices of the Conservative-inspired 'Business in the Community'; provides free and confidential advice to actual and potential very small businesses. It bypasses the local council and is run by an Executive Committee elected from the sponsors of the agency. *Lancashire Enterprises Ltd* was established in 1982 as one of the first generation of Enterprise Boards, and has a considerable presence locally. Unlike some of those Boards, LEL has been concerned with a wide range of sectors, and with providing not only equity capital, but also training, property acquisition and development, cooperatives, and so on. LEL has its premier site in Lancaster, at White Cross, formerly Storeys. By 1986 there were 60–70 users employing about 300 people. About 1 in 7 have been wholly new businesses. LEL, functioning under a Labour county council, has enjoyed a good working relationship with the city council, much better than with some Labour councils elsewhere in Lancashire. This may have resulted from the 'entrepreneurial' zeal of the town clerk and the city architect, and reflects the way in which the 'officers' have exerted a considerable influence over economic initiatives locally. As a local Labour politican argues:

'Well, those who are on the city council . . . say that the councillors do nothing anyway and say that the council is run by the officers. The officer who is actually in charge . . . sees himself as an expert in the whole field of industry and in bringing jobs to Lancaster.' (Bagguley and Shapiro, 1986, p. 34)

There are two main ways of assessing these various initiatives. First, one can compare the total amount spent on 'economic development' and that spent under various headings. There is considerable variation in spending on economic development by different local authorities (Armstrong and Fildes, 1987). Lancaster's total expenditure is close to the median position for non-metropolitan district councils, much less (recently) than Middlesbrough, Knowsley and Thamesdown, but more than Cheltenham and similar to the Isle of Thanet. However, this fairly modest amount is distributed amongst a wide range of activities. For example, the provision of grants and loans is unusual amongst District Councils, yet Lancaster provides between 16 and 25 a year (total value varying between £70,000 and £250,000). It also provides a trade directory, which is done by less than half of such councils. The main limitation is that there is insufficient funding for major land reclamation projects, which may now be a constraint on further development.

Second, one can consider the rate of new firm formation, which varies greatly in different regions and localities (Ganguly, 1984, 1985; *British Business*, 1986). It is possible to analyze VAT registration data for the Lancaster District and compare the stock, starts and stops for

10 different sectors from end-1979 to end-1985. In most sectors the stock of firms registered in 1985 is within 10 per cent of the 1979 figures. The main exceptions are in 'production', which increased from 148 to 201, and 'property, finance and professional services', which increased from 77 to 110. The national changes are similar in direction but show nothing like the same scale of increase. The 35 per cent increase in the number of 'production' firms compares with 8.5 per cent in the north-west and 17.3 per cent in Lancashire (both 1980–4). The increase in 'property, finance and professional services' was mainly caused by extremely low stop rates, that is, such firms have survived well. By contrast the rapid increase in 'production' firms locally was due to extremely high start rates which indicates that local economic initiatives have been successful in generating new manufacturing units. Lancaster District would seem to demonstrate the Thatcherist virtues of vigorous enterprise culture (see Keeble and Wever, 1986, on 'business formation rates' in different regions; Lancaster's is 8.2 production firms per 1000 production employers, compared with 13.8 in the south-east, 8 in the Midlands, 7.1 in the north-west and 4.5 in the north).

Three reservations are necessary: firstly, these new firms will do little to stem the tide of job losses in Lancaster's manufacturing industry (60 per cent of notified job losses 1979–85 were in manufacturing; Lancaster City Council, 1986, p. 5); secondly, since nationally 40 per cent of new firms cease trading within two and a half years the contribution of these in employment terms will be small; and thirdly, there is no sign yet in Lancaster of the development of 'technology-oriented complexes' which, as in Cambridge or Berkshire, seem to be able to provide a set of reinforcing conditions necessary for substantial manufacturing growth.

Conclusion

The Lancaster area has been pulled apart in the last twenty years. In many ways it remains poised in transition between a previous and a future role. The physical signs and many of the people remain from the era of manufacturing industry and the 'family holiday'. The traces of its mid-20th century role in Britain's urban and regional system are still present in its culture and institutional politics, but they are no longer dominant. Economic rhythms are now dictated by state agencies: the next round of restructuring locally will be the story of the logic of state enterprise. The new buildings, jobs in the primary labour market, consumption levels and cultural activities are all heavily dependent on the state. For it has not become an area of

Plate 4.4   Lancaster Castle.

private producer services, nor a favoured location of company headquarters. Rather it is a lower-level regional service centre in northern England, not well endowed with the growth industries of the 1980s.

Still in transition, it has no unified image, no cultural core. On some measures it is comparatively prosperous. House prices increased faster between 1982 and 1986 than in any of the other localities in this study except Swindon, though from a moderately low base. In terms of per capita income, only the population of Cheltenham was better off (see Chapter 6). New firms have sprung up, encouraged by an active local authority able to make something of the assets of a medium-sized, freestanding historic town. Yet at the same time there are signs of social deprivation and economic decline. The unemployment rate is above the national average. Large manufacturing employers have all but disappeared. The construction of Heysham No 2 power station, the principal source of employment in the early 1980s, has just been completed with no sign of alternative jobs. Reductions in public expenditure make work more precarious and less remunerative in a town where the State is the largest employer.

As elsewhere, prosperity and deprivation fall on different shoulders. Aggregate statistics smooth over social divisions, and conceal Lancaster's heterogeneity. Change has meant different things to different social groups. This is encapsulated in the reversal of the roles of the two main urban areas – Lancaster and Morecambe – over the last twenty years. With that transformation have come differential class experiences and a fundamental spatial cleavage.

In Lancaster City itself, deindustrialization has remade the environment. The conservation lobby has ensured that building since the 1970s has been in a neo-vernacular style using local stone. This can be seen in the Lancaster Local Plan where there is a great deal of specification of the sorts of buildings that will be allowed, including the design of multi-storey car parks (both the current car parks are clad in stone). One of the objectives of the plan is to remove 1950s and 1960s buildings and to minimize the destruction of earlier buildings. There is a really detailed planning brief for the new shopping development. Nevertheless this plan has aroused intense public debate. Over 1400 objections were lodged to it within the planning period, and with the current 'hung council' it is certain that further modifications will be made to ensure the compatibility of the development with existing vernacular style (the Planning Department have themselves elaborated the 'Lancaster vernacular').

As industry has departed from the city centre Lancaster has been reconstructed as a modern consumption centre preserving the shells of past rounds of economic structure to house new functions – the old customs house as a maritime museum, the warehouses of the riverfront as gentrified homes, canal-side mill buildings as new pubs. One of the main thrusts of all of this is to construct 'Lancaster' (city) as an object of the tourist gaze. It has many of the ingredients of the modern tourist mecca: a castle (which in 1991 will become entirely usable for 'tourist' purposes); a river and a gentrified river front; the folly-on-the-hill (Ashton Memorial) just restored at a cost of some £1.5 million; four museums, three of which have been recently completed; well-conserved old, interesting streets with 270 listed buildings; cultural events including the Lancaster Literature Festival, and so on. By comparison with, say, Cheltenham, it had none of the atmosphere of a university town in the 1960s. That now has changed because, as Britain becomes in part 'post-industrial', certain kinds of industry and industrial townscapes become central and fascinating objects of the tourist gaze. The physical and cultural attributes of a past industrial heritage have been retained alongside an occupational and industrial distribution dominated by service industry.

White-collar workers living in Lancaster have seen the area transformed in their own image and not surprisingly, they consider it a very pleasant place in which to live and work. They describe an improving environment, with enlarged leisure services such as theatres, restaurants, shops, etc. For such respondents the area is defined in terms of the natural beauty of its surroundings, an attractive historic city surrounded by beautiful countryside with good amenities, while the size of the town contributes to a sense of sociability.

'Oh, Lancaster is not too big, it's not too small, it's close to the countryside ... We've got Morecambe, we've got the seaside, we've got the Lake District, it's a nice place to live.' (clerical worker, LDHA)

However, while almost everyone mentioned the environmental attractions of the area, other groups had rather different experiences. For one manual worker for whom levels of household income had deteriorated in the last five years, 'there's a lot more going on ... [but] we are just getting by'. A part-time nurse commented on the number of restaurants opening up, 'very nice if you can afford them'. Improvements in services were not apparent to everyone and some groups complained about the disappearance of local cinemas and dance halls. Again the experience of the Lancaster area is very different to that of Morecambe. For Morecambe residents a defining feature of the last five years has been a visible decay in the quality of both the built environment and of local services.

The first few decades of the twentieth century saw 'industry' located in Lancaster and 'consumption' in Morecambe, obverse and reverse in a single system of production/consumption. Now industry has departed from Lancaster – for industrial estates beyond the city centre, for green-field sites in the rest of the UK, and for cheaper sites outside the UK. The city is being reconstructed as a post-modern consumption centre, as its occupational and industrial structure has become dominated by service industry and by those with middle class/higher education qualifications. At the same time the built environment increasingly serves to reify the physical and cultural attributes of Lancaster's past industrial heritage.

Morecambe, by contrast, has completely lost its role. Its potential as a centre for a modern tourist industry is extremely limited. Until recently there were no 'conservation areas' in the centre, although many of its commercial buildings are similar to Lancaster's. Its facilities are really rather poor and very unlikely to induce large numbers of people to see it as a site for extended consumption. It does of course possess the 'sea', but for that to be transformed into a regime of pleasure much else would also have to be found. So far it has not been. Indeed, ironically, many of the Morecambe hotels have been kept going because of the influx of construction workers at Heysham 2 (worth £8.5m a year to the local economy) that has saved them, albeit temporarily, from the fate of their counterparts in Thanet. That workforce was, at its peak, equivalent to 31 per cent of the total 1981 male employment in the TTWA. Morecambe has become less a site for consumption, relative to its past and to other places, and more a centre for certain kinds of industry (nuclear, gas and possibly oil). Morecambe's lack of a strong professional middle class has meant that there has been almost no opposition to these developments – indeed they have been enthusiastically embraced. Also, many of the hotels and guest houses have been-converted into accommodation for DHSS claimants, for students at the University, and for elderly people. So, although Morecambe is dominated by the 'politics of tourism', its social composition is more like an inner city area. One might almost describe it as Lancaster's 'inner city', especially in the light of its housing conditions. In many respects it is Morecambe that bears the greatest costs of the recent round of restructuring.

The spatial impact of restructuring has, then, been profound even at intra-locality level. Neighbourhoods have been transformed, their trajectories of development realigned. Visible in many spheres of local life – in housing, in the built environment, in economic life, in standards of living – the tendency to spatial reversal is most neatly encapsulated in changing patterns of political behaviour. On the one hand there is the engineer quoted above who, when he came to live in Morecambe and work for the CEGB at Heysham, felt compelled to enliven the Labour Party. On the other hand there is the emergent territorial dimension in local politics after Local Government Reorganization as councillors, irrespective of party, lined up as defenders of Lancaster or Morecambe.

References

Armour, Wendy, Mark-Lawson, Jane, and Walby, Sylvia, (1987) 'Gender politics in Lancaster', Working Paper No. 31, Lancaster Regionalism Group, University of Lancaster.

Armstrong, Harvey and Fildes, J. (1987) 'District council industrial develop-ment initiatives and regional industrial policy', *Progress in Planning*, Vol. 28.

Bagguley, Paul (1986a) 'Economic restructuring and employment change in Lancaster 1971–1981: manufacturing industries', Working Paper No. 19, Lancaster Regionalism Group, University of Lancaster.

Bagguley, Paul (1986b) 'Service employment and economic restructuring in Lancaster, 1971–1981', Working Paper No. 20, Lancaster Regionalism Group, University of Lancaster.

Bagguley, Paul (1987) 'Flexibility, restructuring and gender: changing employment in Britain's hotels', Working Paper No. 24, Lancaster Regionalism Group, University of Lancaster.

Bagguley, Paul and Shapiro, Dan (1986) 'Lancaster; diversity in decline?' Paper to the Conference of the Institute of British Geographers, Reading, January.

Bell, C. (1968) *Middle Class Families*, London: Routledge and Kegan Paul.

British Business (1986) 'UK registrations and deregistrations for VAT', 19 September, pp. 6–7.

Fogerty, M. P. (1945) *Prospects of the Industrial Areas of Great Britain*, London: Methuen.

Fothergill, S., and Gudgin, C. (1979) 'Regional employment change: a subregional explanation', *Progress in Planning*, 12, pp. 155–220.

Fothergill, S., and Gudgin, G., Kitson, M. and Monk, G. (1986) 'The de-industrialization of the city', in R. Martin and B. Rowthorn (eds), *The Geography of De-industrialization*, London: Macmillan, pp. 214–37.

Fulcher, M. N., Rhodes, J. and Taylor, J. (1966) 'The economy of the Lancaster sub-region', OP10, Department of Economics, University of Lancaster.

Ganguly, P. (1984) 'Business starts and stops: regional analyses by turnover size and sector, 1980–3', *British Business*, 2 November, pp. 350–3.

Ganguly, P. (1985) 'Business starts and stops: UK county analysis, 1980–3,' *British Business*, 18 January, pp. 106–7.

Keeble, D. and Wever, E. (1986) 'New technology and high-technology industry in the United Kingdom: the case of computer electronics', in D. Keeble and E. Wever (eds) *New Firms and Regional Development in Europe*, London: Croom Helm, pp. 75–104.

Lancaster City Council (1986) *Lancaster District: A Summary of Recent Employment Trends and Prospects*, Lancaster: Lancaster City Council.

Lancaster Guardian various dates.

Littler, Craig (1982) *The Development of the Labour Process in Capitalist Societies*, London: Heinemann.

Lloyd, P. and Shutt, J. (1985) 'Recession and restructuring in the north-west region, 1975–82: the implications of recent events', in D. Massey and R. Meegan (eds): *Politics and Method*, London: Methuen, pp. 16–60.

Mark-Lawson, Jane, Savage, Mike and Warde, Alan (1985) 'Gender and local politics: struggles over welfare policies, 1918–1939', Lancaster Regionalism Group, *Localities, Class and Gender*, London: Pion.

Martin, P. (1986) 'Thatcherism and Britain's industrial landscape', in R. Martin and B. Rowthorn (eds), *The Geography of De-industrialization*, London: Macmillan, pp. 238–91.

Martin, Roderick and Fryer, Bob (1973) *Redundancy and Paternalist Capitalism*, Hemel Hempstead, Herts: George Allen and Unwin.

McClintock, M. (1974) *The Quest for Innovation*, University of Lancaster.

Murgatroyd, Linda (1981) 'De-industrialization in Lancaster: a review of the changing structure of employment in the Lancaster district', Working Paper No. 1, Lancaster Regionalism Group, University of Lancaster.

Murgatroyd, Linda and Urry, John (1985) 'The class and gender restructuring of the Lancaster economy, 1950–1980', Lancaster Regionalism Group, *Localities, Class and Gender*, London: Pion.

NEDO (1986) *Changing Working Patterns*, London: National Economic Development Office.

Piepe, A., et al., (1969) 'The location of the proletariat and deferential worker', *Sociology*, 3, 2, pp. 239–244.

Piore, M. J. and Sabel, C. F. (1984) *The Second Industrial Divide; Possibilities for Prosperity*, New York: Basic Books Inc.

Rustin, M. (1986) 'The fall and rise of public space', *Dissent*, Fall 1986, pp. 486–94.

Savage, Mike (1987) 'Spatial mobility and the professional labour market: a case study of employers in Slough', mimeo, Urban and Regional Studies, University of Sussex.

Shapiro, Dan (1988a) 'Social change and the Lancaster labour market 1971–1981', Working Paper No. 34, Lancaster Regionalism Group, University of Lancaster.

Shapiro, Dan (1988b) 'The state and restructuring' in J. Morris, A. Thompson and A. Davies (eds), *Labour Market Responses to Industrial Restructuring and Technological Change*, Brighton: Wheatsheaf.

Tolliday, S. and Zeitlin, J. (eds), *The Automobile Industry and its Workers: Between Fordism and Flexibility*, Cambridge: Polity Press.

Urry, John (1987a) 'Economic planning and policy in the Lancaster district', Working Paper No. 21, Lancaster Regionalism Group, University of Lancaster.

Urry, John (1987b) 'Holidaymaking, cultural change and the seaside', Working Paper No. 22, Lancaster Regionalism Group, University of Lancaster.

Warde, Alan (1988) 'Conditions of dependence: the roots of working class quiescence in Lancaster', Working Paper No. 30, Lancaster Regionalism Group, University of Lancaster.

5

The Isle of Thanet: Restructuring and Municipal Conservatism

NICK BUCK, IAN GORDON,
CHRIS PICKVANCE, & PETER TAYLOR-GOOBY[1]

Introduction

The 'Isle of Thanet' has not really been an island since the silting up of
the Wantsum Channel during the Middle Ages, but it remains quite
isolated in its corner of north-east Kent. Nowadays the only sea to be
crossed in approaching Thanet from the west is a flat expanse of
unfenced cabbage fields but this is sufficient to underscore a local
sense of separateness, on occasions even of rejection. Kent County
Council is thought to be run by representatives from the commuter
belt of west Kent; a series of reorganizations in the public sector have
removed administrative responsibilities from Thanet; and the recent
strengthening of Canterbury's role as the sub-regional shopping
centre has enhanced feelings of relative deprivation. Although in
other ways quite different, the somewhat insular mentality of Thanet
people is akin to that noted on the Isle of Sheppey, 30 miles further up
the Thames estuary (Pahl, 1985). And the area along this Kent coast is
exceptional within a generally prosperous region in terms of its high
unemployment and pockets of real deprivation. With over 100,000
residents (and 7000 unemployed) Thanet has the most substantial
problems–and its unemployment rate (of 17.4 per cent in August
1987) is above that of any other travel-to-work area within 200 miles.

In fact, Thanet stands out from the rest of Kent on a whole series of
social indicators, including juvenile crime, single parent families,
dependence on benefits, and car ownership levels – a complex of
problems which a report by local teachers saw as involving a
distinctive 'Thanet dimension' (Martin and Ellis, 1982). Alongside
these real problems there is a tendency to 'moral panic' linked to
perceived threats to the resort image from invasions particularly of
young people. In the 1960s this was focused on the Mods and Rockers
who came and fought on Bank Holiday weekends – the 'petty

[1] Urban and Regional Studies Unit, University of Kent at Canterbury

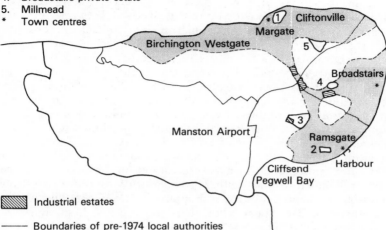

THE ISLE OF THANET

Interview study areas:
1. Central Margate
2. Ramsgate
3. Newington
4. Broadstairs private estate
5. Millmead
* Town centres

Cliftonville
Margate
Birchington Westgate
Broadstairs
Manston Airport
Ramsgate
Harbour
Cliffsend
Pegwell Bay

Industrial estates

——— Boundaries of pre-1974 local authorities

- - - - Extent of built-up area

Figure 5.1 The Isle of Thanet.

sawdust Caesars' castigated by a local magistrate; in the 1980s it focused more on the young migrant unemployed seeking cheap seaside accommodation and casual employment, who were seen as the harbingers of a Costa del Dole. At the same time it is clear that problems are by no means the monopoly of 'outsiders'. A large part of the local population, particularly among the retired and older workers, are themselves migrants from elsewhere.

Thanet's modern identity is the creation of the 1974 local government reorganization which merged the administration of the three distinct resort towns of Margate, Ramsgate and Broadstairs, and their rural hinterland. In political terms they have not been integrated fully, however; each retains its own particular image and market niche in the tourist and retirement trades, and each has somewhat different clusters of economic interest. Margate, although one of the earliest seaside resorts of the gentry, has for most of its history specialized in a more cheerful, vulgar or 'Cockneyfied' style of tourism, epitomized in the Dreamland amusement park on its Golden Mile fronting the well used sands. Ramsgate, to the south, shares some of Margate's character, although with fewer day trippers, but its principal feature is its substantial harbour, catering both for yachtsmen, and less picturesquely for commercial traffic including both Volkswagen imports and the Sally Line's ferry operation to Dunkirk. Industrial activities have also assumed rather more importance in the town than elsewhere in Thanet. Finally, there is Broadstairs, another Victorian resort, wedged between the two larger towns but still retaining a much quieter character and considerable charm, with Charles Dickens's Bleak House its nearest counterpart to Dreamland. What unites the three towns, apart from the substantial travel to work between them, is the fact that tourism to each of them has been in decline since the 1950s and needs either regeneration or replacement by some other economic base. The economic interests of those actively involved in local politics remain somewhat distinct, however, and this has contributed to the fragmented and factional character of policy making in Thanet. The determination of each town to maintain its shopping centre has meant that some multiple stores have not located in Thanet at all since Thanet Council would neither give priority to one town over another, nor allow out of town shopping development. This helped the growth of Canterbury where there were no such hesitations.

In many respects Thanet is less peculiar than its situation, its history or its self-perceptions indicate. Indeed, although it has been indelibly marked by its history as a resort area, its economic activities are now more diverse, including (up to the recession) a fairly average proportion of manufacturing employment. In social terms, too, it is much less uniformly middle class than either the resort image or its continuity of Conservative representation at Westminster would suggest. Even its residential geography has recognizable parallels with larger cities. Each of the three towns has a central area close to the sea front with a mobile and socially deprived popualtion, surrounded by rings of subsequent development which are progressively more affluent, until one reaches the areas developed in the post war period, which include both comfortable owner occupied neighbourhoods and peripheral council estates with some of the most disadvantaged households in Thanet.

To give some flavour to the differences in social struture which exist within Thanet, and to emphasize the importance of not treating such localities as uniform entities with common problems and interests, we give brief sketches of six areas, five of which were studied as part of our household interview survey. The areas were also identified from an analysis of demographic and social data, and represent most of the major types of residential neigbourhoods to be found in Thanet.

Central Margate is one of the oldest areas and some at least of its current physical fabric dates back to the eighteenth century. However much of what now exists is a product of very piecemeal tourist-related development and redevelopment in the nineteenth and twentieth centuries. The front contains a mixture of amusement arcades, a few hotels and largely vacant shops abutting a busy road, and then a sandy beach, still very crowded on summer weekends. Behind are a few streets of considerable character, an unremarkable 1960s shopping development and, to the east, areas of dereliction and disrepair. Socially the area is very mixed indeed, but it contains relatively few families with children. The considerable amount of rented housing attracts a rather mobile population. The most extreme instance is the growing use of hotel and boarding house accommodation for long-term residents, particularly those in receipt of social security payments. These included young single people, until changes in benefit regulations interrupted this, and still include many unemployed families and older single people as well as those discharged from mental hospitals. The stock of relatively cheap low-grade accommodation of various types in the area thus mitigates the housing problems of a much wider region. The result is a major local concentration of unemployed, with current unemployment rates probably well over 40 per cent. This area also contains considerable numbers of self-employed people who have set up businesses, particularly shops and hotels. Many of these people came from outside Thanet, again because Thanet provided cheap property. Few appear to have succeeded in any large way, and many have failed and moved on, some to other parts of Thanet. However they make up a large part of the population of the area at any time.

Our second area was part of the inner Victorian and Edwardian suburbs on the north side of Ramsgate. The area consists of terraced houses, some of which are three and four storeys, and are let as flats or rooms, and others smaller houses, mainly in owner occupation. Socially this is very much a working class area, and many of the people we interviewed there worked, or had worked, in manufacturing jobs. But, like the Margate area, it was characterized by a rather high turnover of population, and there were many single and divorced people and single parent families. Like central Margate, this was an area of relatively cheap and accessible housing opportunities. People had moved between jobs frequently, many had made considerable use of seasonal jobs and jobs in the tourist industry, and many had experienced lengthy spells of unemployment.

Newington, also in Ramsgate, is a council estate built in the 1950s. It was built largely as a cottage development with considerable attention to street layout and open space, but now looks uncared for and conveys a sense of dereliction. Like most council estates of this type it has a very stable population, and the current residential mix reflects the age of the estate. A 'family' household structure predominates, with more children than the central areas. In some ways this represents the most stable working class area of Thanet, and the level of job stability is higher here for people in manual jobs than elsewhere. Also, more people work in manufacturing here. This area has the highest level of Labour voting and, from our survey, a greater tendency to use the language of class in interpreting local and national problems. Like other working class areas of Thanet, Newington is characterized by high levels of unemployment and what seems an exceptionally high level of long-term sickness.

Birchington, which is on the north coast five miles from Margate, began to grow in the nineteenth century but has seen substantially more growth in the post war period. This has mainly been in the form of housing for the elderly, including many bungalows as well as larger houses, and like other parts of the coast west of Margate it has been one of the major destinations for retirement migration. Although now part of a continuous development along the coast it still strives to maintain a separate 'village' identity. The proportion of elderly in the population in 1981 was 40 per cent. Birchington is relatively affluent compared with the rest of Thanet, and political interests here (and in other retirement areas of Thanet) tend to diverge from the tourism interest: the main issues in Birchington relate to low rates and resisting intrusive development.[2]

Our last two areas were both built up in the 1970s but they offer extreme contrasts in social disadvantage. The first is a private owner-occupied estate on the outskirts of Broadstairs with an open plan design which might be found in any small town development of this period. In Thanet terms this area has a relatively affluent and relatively young population. There is a greater contrast than elsewhere between the work experience of men and women, largely because the men's work histories are so different from the Thanet norm. They are generally in stable, skilled, working class or non-manual jobs, and there are some self-employed people, while the women's work experience is much more irregular, with few longer term jobs (those there are being largely public sector) and, as elsewhere, a reliance on short term catering or tourist-related jobs. However, even in this area some men have experienced redundancies and there are unemployed teenage children.

Millmead, finally, is a 1970s council estate on the southern outskirts of Margate, though parts of it were in fact built by a speculative private builder, so that its building style is not so distinct from the previous area. It is however a classic 'sink estate', with very high proportions of single parent families and long-term unemployed, and

[2] Because our concerns relate principally to the working age population of Thanet, Birchington was not included in our interview survey.

it has the worst reputation of any area of Thanet, though this seems little justified by its appearance, which is remarkably free of signs of vandalism or graffiti. In the past both men and women moved frequently between jobs, but as unemployment increased more seem to be in very long-term unemployment. Women appeared to be more likely to be employed than men, but mainly in low-paid service jobs, such as cleaning, or in very occasional jobs. Our survey suggested that many people here were largely alienated from formal politics of any kind.

'Locality' in Thanet thus obviously means several quite different things. At one level the whole 'island' is a distinctive locality with an economy, labour market and sets of political expectations shaped by its particular history of economic development, and ultimately by its peripheral location. At this level it faces a serious problem of long-term decline in its economic base and a set of strategic questions as to how (and if) its restructuring can be successfully managed. For other purposes, particularly for the representation of economic interests, and for local party organization, the effective localities have been the individual resort towns. The sectoral balance of interests is different in each, and this has formed the basis of cleavages in local politics. A further factor is that in attracting visitors, and retirement migrants, it is the image of the specific towns which are felt to be of critical importance. Hence a major interest in each is the preservation of the local environment and the upgrading of its particular attractions. But for perhaps the majority of Thanet residents, particularly for those not currently working, the critical locality is that of the residential neighbourhood. In an area where few people work in stable jobs in large establishments, and many are inactive, social interaction on a residential level is likely to play a more substantial role than elsewhere in the development and reproduction of local culture. It is only at this level that class differences within Thanet become apparent – with the corollary that groups such as the retired community in Birchington can remain in substantial ignorance of the real conditions of life of, for example, Millmead residents. And it is in the distinctive social situations of a number of these groups, as much as in the general economic decline, that Thanet's current problems lie.

The state of Thanet in the late 1980s, and the prospects for its regeneration, depend on the interaction of economic, social and political factors, all heavily influenced by its pattern of past development. In the remainder of this chapter we shall consider the three main elements in its recent history – looking in turn at the restructuring of its economy through several rounds of investment, at labour market processes and outcomes, and at local council economic intervention and local politics.

Three rounds of economic restructuring in Thanet

The current economic, social and political structure of any locality depends on the history of rounds of investment in the past and the residue they have left: in the built environment and the currently surviving businesses, but also in the practices of the labour market and of local politics. Moreover, this history will profoundly affect the way that a locality experiences and modifies current processes of economic change.

We would see Thanet in terms of three such overlapping phases: its period as a tourist resort, the period from the 1950s to the 1970s when its role as a branch plant manufacturing economy grew as tourism went into decline, and the latest period after manufacturing went into decline and Thanet sought a variety of roles, including private sector welfare provision and port expansion.

The growth and decline of tourism

The three Thanet towns – Margate, Ramsgate and Broadstairs – largely owe their transformation from small towns based on agriculture, fishing and shipping to the growth of tourism, dating back to the eighteenth century. Indeed Margate not only lays claim to being the first resort to develop sea bathing, during the 1730s, but also has some claim to being the first resort to cater for a mass market, rather than simply the upper class. This was based on its relatively cheap water communications from London, developed in the latter part of the eighteenth century.

The major growth of seaside resort towns, however, occurred in the nineteenth century. Their growth was linked to the spread of new forms of leisure for the residents of the growing industrial towns and cities: mass holidaymaking and day excursions, following the mechanization of transport. The development of the steamboat service and then a direct rail connection from London in the 1860s gave impetus to a fresh period of growth, and the populations of the three Thanet towns, which had been 13,000 in 1801, grew from 32,000 in 1861 to 78,000 in 1911. The three towns developed in rather different ways, with Margate gaining an increasingly working class image, particularly because of the importance of the day tripper from London. The other two towns had a somewhat 'quieter' image, with Broadstairs the more exclusive. Even Margate had its share of larger and more expensive hotels, particularly in Cliftonville – the name itself being designed to detach the area from the image of Margate. Tourist-based growth continued through the interwar period in spite of the Depression of the 1930s. The three town councils, and

Plate 5.1   The Royal Sea Bathing Institution, Margate.

particularly Ramsgate's, were heavily involved in continued investment in tourist infrastructure, building promenades and swimming
baths, and laying out parks, and there was also major private
investment in the Dreamland amusement park in Margate (Stafford
and Yates, 1985, p. 121). The scale of tourism in the local economy of
Margate can be gauged by figures for 1939, showing 30 large hotels,
60 smaller hotels, 150 private hotels, 1300 boarding houses, 1500
apartment houses and 3500 private houses catering for visitors. Thus
6540 of the 14,000 rated properties in Margate were used for
accommodating tourists at some part of the year (Pimlott, 1947). The
dominance of small, low-cost forms of accommodation is also
striking. This was very much an area of cheap holidays in contrast to
resorts serving more affluent parts of the market, based on larger
hotels, such as Bournemouth. Together the Thanet resorts constituted the largest tourist destination in the south-east. Before the
Second World War there was very little diversification away from
tourism, with hardly any manufacturing; most industries serviced the
local population or the tourists.

After the war, which itself had a drastic effect on the towns, tourist
activity was sustained into the 1950s but this depended on extending
the catchment area, particularly into the Midlands. In the immediate
postwar period there was considerable optimism about the prospects

of seaside resorts, based on the extension of paid holidays, but other major changes in patterns of tourism overtook this. Expanding car ownership increased the range of opportunities, and the major areas of tourist growth in Britain were the countryside or more remote coastal areas, particularly in the south-west. The cities which had been the source of major flows of the tourists to traditional seaside resorts, supported by both custom and organized transport links, were themselves in decline. Changing consumption and living patterns changed the desired content of holidays (see Urry, 1988); and finally large tour companies and cheap air travel expanded the opportunities for overseas holidays, particularly in the low wage areas of the Mediterranean.

Thanet probably experienced the restructuring of the tourist industry more painfully than any other large resort in Britain. In large part this reflected its down-market image and its dependence on the most traditional form of holiday, the seaside hotel and boarding house. Initially the Thanet towns lost their more affluent customers, for whom more opportunities were arising, and this led in the 1950s to the closure of many of the larger hotels. In this period, though, the aggregate level of visitors probably did not decline. During the 1960s, however, nearly half of all hotel rooms were lost and this decline continued through the 1970s, with heavy employment losses in hotels. One index of the decline is the number of hotels listed in the guides. In Margate in 1950 there were 571, a level which was broadly held until 1961. By 1971 the number had declined to 407, though in the 1960s there was a growth of self-catering flats. However both categories declined sharply in the 1970s and 1980s, and by 1985 only 124 hotels were listed. Moreover, the distribution of visitors was changing. The proportion of staying visitors aged over 65 rose from 17 per cent in 1967 to 47 per cent in 1984. The resorts were increasingly relying on the return visits of those who had been coming regularly over the years, and few new visitors were being attracted. The small scale of most hotel operations and the market segment they aim at has resulted in the failure of Thanet to maintain its conference status – an area where other resorts have invested heavily. Thanet has been increasingly unable to offer the quality of hotel accommodation required. The nature of the hotel stock has also led to difficulties in meeting growing demands for more bathrooms and for central heating, and the industry's weak capacity to restructure is reflected in two revealing statistics: few English Tourist Board Section 4 grants for upgrading are taken up, and most graduates from Thanet Technical College in hotel and catering management are not attracted by the opportunities in Thanet and leave the area.

The mechanisms through which restructuring processes affect

local tourism diverge strongly from those in manufacturing on which previous research has concentrated. A first reason for this is the dominance of very small firms, or self-employed workers, in the tourist trade. There has been very little involvement of large capital in running tourist activities in seaside resorts – larger hotel chains, for example, are very little represented in the resorts. The large hotels which do exist tend to be in individual ownership, or at most are owned in small groups. In resorts such as Thanet the hotel industry has largely consisted of very small hotels which employ few people outside the owner's family, and sometimes open for only part of the year. The Hoteliers Associations, representing such hotels, are increasingly unable in a period of decline to stem the growth of what they see as unfair competition from guest houses, or the drift towards uses of hotels which undermine the resort, such as long-term accommodation for unemployed single people.

The second factor is the significance of locality and the built environment for the product being sold. Tourism is a resource-based industry, but one where the resource has to be maintained, regulated and developed in the face of changing external demands. In an industry of small firms this puts considerable stress on the role of the local authority in preserving the image of the resort, promoting it outside, and extending the facilities available, since individual firms are unable to perform these roles. This leads to a characteristic style of local politics and local council intervention which we label 'municipal conservatism', as a mirror image to municipal socialism.

This has had five facets. Firstly, the councils have devoted particularly high levels of expenditure to physical maintenance and providing environmental services such as street cleaning, upkeep of roads, and open spaces. Secondly, the local authority normally took responsibility for advertising and marketing the resort outside the area. This was a major area of confusion after the creation of Thanet District Council, with opposition from the three towns to Thanet projecting a combined image. Thirdly, it ran major tourist facilities such as theatres, parks and beaches, often sustaining losses as we shall show when discussing the Margate Winter Gardens. Fourthly, there was a role of regulation and social control. Planning machinery has been used to preserve tourist uses in certain areas and prevent conflicting uses, as in the Margate holiday zone. Local authority by-laws were used as an instrument to control the behaviour of visitors. Finally local authorities could have a role in making major strategic decisions about how to adapt to changes in the pattern of tourist demand – by changing or reinforcing the image of resort. For example, in the 1960s and 1970s a number of British resorts made heavy investments in conference facilities or in yachting marinas.

The third distinctive feature of the restructuring of tourism relates to the role of larger capital. While it was generally not involved in the provision of tourist services, it has occupied a new role as a broker for tourist services. The tour companies increasingly take over the role of advertising resorts and putting together packages of transport and accommodation services. A division has thus emerged between large capital which is not tied to particular localities, and small capital which is. Localities throughout Europe thus compete to sell their facilities to companies which can use this competition to force down prices. Some localities may be left out altogether. Even within Britain seaside resorts have come to depend on the brokerage role of coach companies. It thus becomes more difficult for resorts such as Thanet to bring themselves to the attention of potential customers. Increasingly in Britain the creation of the required image depends on very heavy investment in new leisure facilities, such as theme parks or marinas, which again leaves the resorts dependent on non-local large capital.

Attracting manufacturing industry: 1950–1970

Policy makers and politicians had been aware since the war that tourism provided an inadequate base for the economy, leading to high winter unemployment. However the government view was that moving in industry was not an appropriate solution to the problem. A 1948 white paper on 'The Distribution of Industry' argued that

the problem of the Holiday Areas is not one suitable for solution by scheduling as Development Areas. Factory buildings on any scale might interfere with the Holiday Industry of the town, which with holidays with pay, should become still more active in these areas. The best help the Board of Trade could give would be to encourage industries having a seasonal demand for labour in winter to go to these areas, but such industries are few and far between (quoted in KCC, 1952, p. 39)

By the late 1950s and early 1960s it was clear that hopes of growth were not going to be realized, and that instead Thanet tourism was going into decline. The local authorities became increasingly involved in trying to attract manufacturing industry to provide an alternative base for the economy. This included both lobbying national government and involvement in estate development. Thanet's relatively successful development of a manufacturing sector depended on 'interventionist' local government. The support for industrial growth did involve some conflict within the local authorities with tourist interest, due to fears that it would lead to increased

Plate 5.2   Hayne industrial estate, Ramsgate.

wage rates and labour shortages, and that industrial development would detract from the image of the places, though in fact sites for industry were developed well away from the seaside.

In the short term some success was achieved: in the 1960s and 1970s, a considerable amount of mobile industry was leaving the larger cities in search of lower wages and uncongested sites. Thanet already had a flexible workforce accustomed to low wages, and a programme of industrial estate development succeeded in accommodating a good number of firms. Many firms moved from London from the late 1950s onwards, and from a number of other high-wage or more unionized areas. There was a brief period of Assisted Area Status (1958–1961) and a subsequent permissive regime for Industrial Development Certificates, but it is not clear how much of the very high growth of the early 1960s follows directly from this, and how much from other factors. The result of this growth was that the share of manufacturing in employment rose from a little over 10 per cent in the late 1940s to almost 30 per cent in the early 1970s.

The firms which came to Thanet certainly reflected the conditions under which this growth took place. They were drawn from a wide range of sectors with no particular industrial specialization. Products included transformers (GEC), marine radar (Racal), toys (Hornby – the largest manufacturing employer), office furniture, radiators, printing, clothing and plastics. Our interviews with firms suggested than none had strong purchase or supply linkages with any other local firm. Firms included many branch plants undertaking routine assembly work. When local design or marketing occurred it was for largely standardized products. The single plant firms were again mainly producing products with unsophisticated technology. Many such firms were acquired by larger companies in the 1960s and 1970s. The bulk of the work was routine assembly and there was little demand for specialized labour. There was no sign of firms locating in Thanet because of its proximity to Europe, and none of the firms studied in detail anticipated any benefits from the Channel Tunnel.

Thanet thus developed as a peripheral manufacturing location with no very distinctive benefits. Its major advantage, a plentiful supply of low-wage labour, was one it shared with many other places at this time. Interestingly, in this context, the only managers from different firms who meet on a regular basis are the personnel managers to compare wage rates and discuss other employment issues. The new manufacturing did not inject stability into the Thanet labour market as had been intended. Nor did it provide much in the way of training to generate a pool of more skilled workers which might attract other firms. Indeed difficulty in recruiting the small numbers of skilled workers required, such as maintenance engineers, was one of the major problems reported by Thanet firms. The managers generally stressed the scale of the black economy and the problems of a culture of unemployment. In consequence some firms with more stable levels of production perceived a problem with the instability of the workforce. In the past they lost workers to summer jobs in tourism, but now agriculture gave rise to more of a problem. One personnel manager referred to younger workers, 'who haven't got the family commitments' while in contrast 'older blokes tend to watch their Ps and Qs. They have been made redundant many times.' Another manager from a firm with much more unstable employment referred to 'an endless supply of the type of people we want'. However the single most frequently mentioned problem was the isolation of Thanet, in terms of its distance from most British markets and from central industrial regions. There was also a more metaphysical sense of isolation from industrial Britain. 'The place has an inbuilt lethargy. It is almost falling off the globe here. There is a temptation to look out to sea.'

Thanet has had little direct experience of the more recent patterns of *in situ* restructuring in Britain such as the use of new production process technology or more flexible approaches to products or markets, though there has been greater use of temporary contract labour. Instead it has suffered from numerous closures through rationalization. Job gains of the period of manufacturing investment have proved extremely fragile and in the recession from the late 1970s onwards the area has experienced a very high rate of closures, in particular as the large multi-plant firms reconcentrated on more central sites. The rate of manufacturing job loss has been particularly high since 1981 – a decline of 22 per cent between 1981 and 1984. Within the year to August 1987, Racal, GEC and Associated British Foods, along with a number of other single plant firms, have closed factories in Thanet.

Beyond manufacturing: looking for a new 'round of investment'

The restructuring processes of the 1960s had been based on the view that for most industrial activities location had ceased to matter greatly. There were some areas, above all the large cities, where conditions were perceived to be particularly adverse, but for the rest the availability of land and labour with no particular distinguishing qualities was sufficient. In this context Thanet, with a little help from central government, had benefited from substantial industrial growth. After the mid-1970s recession all this began to change. Nationally, production activities were in decline, and those that remained were much more likely to be exploiting particular market niches with more specific locational requirements. Instead services, particularly financial and business–related services, and some private consumer services, were accounting for most of the employment growth. Growth of this kind tended to follow the regional and subregional hierarchy of service centres, and Thanet conspicuously failed to gain in this process.

Instead, the last decade has been characterized by a desperate search to identify and exploit some specific locational advantage of Thanet. This has involved the local council in a long series of attempts to attract large capital to exploit some perceived opportunity in the area, including attempts to develop a major tourist project, particularly in Ramsgate or Pegwell Bay; or, following the closure of the Hoverport, to promote Manston aerodrome as an international airport for Kent; or to sustain a cross-Channel ferry service from Ramsgate. These projects have been characterized to date either by a complete absence of progress or by a combination of excessive aspirations and very modest achievements. There are, moreover, intrinsic conflicts between some of them.

All these projects have to be set against Thanet's unattractiveness for capital investment in most industries, and especially the advanced ones. This stems partly from geographical circumstances, but is largely due to the responses made to these over the last century. It is thus a self-reinforcing historical process. At a national level, the area

179

is unattractive for manufacturing: its accessibility to markets is very poor due to the inadequate road and rail links, and its environment, while attractive to older migrants on relatively low incomes, does not compete with that of Berkshire or other parts of Kent. Its low-paid and compliant labour force is not distinctive to it, and it suffers a great shortage of qualified workers. Thus we find that among the industries where Thanet has had some recent success the Sally Line cross-Channel ferry operation has depended considerably on financial support from the local authority, while the growth of private nursing or old people's homes has been based on the re-use of premises vacated by the holiday industry. The Council's role in relation to the Sally Line will be discussed below: here we shall focus on the residential homes business.

The growth of the *private welfare sector* needs to be seen in the context of the increased percentage of elderly people in the population as a whole. Like many resort towns, Thanet has a substantially above average share of elderly in its population. Those of retirement age made up 28 per cent compared with under 18 per cent nationally (1981 Census) and more than 10 per cent of the population was aged over 75. This has been primarily due to retirement migration. The growth of home ownership has given retired people much more opportunity for mobility and the consequence has been a substantial elderly migration to seaside resort towns. The other cause of an increasing dependent population has been the policy of closing geriatric and psychiatric hospitals. Thanet has gained ex-patients (or 'become a dumping ground') following hospital closures throughout Kent.

The growth of the private welfare sector has also been influenced by two other policies: the shortfall of public health and social service provision, and a deliberate policy of encouraging private sector health and social service provision.

Social services provision falls far short of national guidelines. A meals on wheels service is provided at 9 per cent of the recommended level, home helps at 27 per cent, residential provision at 30 per cent and day centre provision at 64 per cent of DHSS norms. And in health, resources for the Canterbury and Thanet district as a whole stood at about 85 per cent of the level required for parity with other districts in the region (taking into account differences in mortality and age structure) (CTDHA, 1985, DI). Geriatric bed provision is at about 90 per cent of the regional norm and spending on mentally ill and handicapped people about three-quarters of the parity level. Canterbury contains about twice as many acute and maternity hospital beds and rather more geriatric beds than Thanet, although Thanet's retired population is about one and a half times that of

Plate 5.3 'No vacancies' in January: small hotels in Margate.

Canterbury. The shortfall in public provision is made up by the private sector.

The encouragement of the private welfare sector is a boost to a long-term trend. Thus between 1971 and 1981 the number of beds in private homes for the old and disabled in Margate and Ramsgate increased from 900 to over 1500. This is partly linked to the resources available to old people through selling their homes. However a major boost was given to private residential provision by the DHSS policy, introduced in 1980, of subsidizing fees for those who fall within its means-test limits – though no test of need for residential care is applied. In Thanet the number of people resident in old person's homes has risen from 1654 in the 1981 census to just over 2000 in a 1985 survey (Community Health Council, 1986). This amounts to a major new industry with a turnover estimated in excess of £20 million. The local social security office estimates that some 1200 residents are assisted. This would imply a DHSS contribution of £7.5 million.

Thanet has also been the test site for an innovative community care experiment. This relies on untrained local volunteers (usually neighbours) to provide care to people living in their own homes, in addition to that provided by peripatetic social services staff. They are paid on a fee for service basis. Ninety-four per cent of carers are women and many have experience of care work (about half have worked as care assistants, a quarter as nurses and a fifth in social services (Challis, 1984, pp. 363–4). Community care offers a cheap alternative to the provision of council residential accommodation since it means that fewer trained social services staff receiving normal wages are employed. It is perhaps significant that it has been tried in Thanet rather than Canterbury.

The rundown of public sector residential accommodation, coupled with the expansion of private provision and community care, have had two labour market effects. Firstly, private provision offers opportunities for entrepreneurship in a relatively secure environment due to the state-subsidized fees and high demand. The private homes listed in the Health Council survey are mainly run by owner/manager/matrons with one establishment, although a local Conservative councillor runs five, accounting for over 300 beds, or a sixth of those on Thanet. As other studies show, the sector provides oportunities for female entrepreneurship, a third of the owner-run homes being run by women (Judge, 1985). Only five of the homes are run or owned by medically qualified people (four SRNs and a doctor who apparently advertises his home in his own GP waiting room) although many others are likely to have relevant experience as unqualified health authority or social service care workers.

Secondly there is a considerable demand for routine domestic workers in residential homes and care workers in the community, at relatively low levels of pay. The main employment effects of the policies applied in Thanet have thus been on women – both as employees of the NHS and social services department, which have been subject to substantial employment losses from 1983 onwards, and as the main providers of non-state private and voluntary care.

Whilst many people have been involved in the provision of care services, and the combination of a low level of state provision, state-supported privatization and the attraction to Thanet by cheap accommodation of many welfare dependents adds something to the local economy, it does little to replace the previous economic bases of Thanet.

What now supports the economy of Thanet, in terms of activities which draw in income from outside, can be divided into three components. The first consists of the remains of tourism, which draws visitors to Thanet. The second component consists of the production of goods and services consumed outside the area, that is to say by agriculture, manufacturing and the cross-Channel transport services. Finally there are the incomes of residents derived from outside the area, including those who commute out to work, and those who have retired to Thanet. A very approximate calculation suggests that only around 20 per cent of local employment now depends directly or indirectly on tourism, while 50 per cent depends on the second group of activities, and the remaining 30 per cent can be attributed to the demands of outward commuters and retirement migrants (calculated as the excess proportion of the retired population over the national average). While tourism is now only a small element of the economic base, production activities are not overwhelmingly dominant. Looked at another way, the calculation suggests that around half the economic base still depends on the environmental attractions of Thanet to visitors or residents. We suggest later that the political implications of this sort of split have hampered some attempts to restructure the economy of Thanet.

The Thanet labour market

Thanet's central problem, and its main claim to assistance, is its high level of unemployment, involving currently almost one in five of its working population, with unemployment rates in recent years comparable to those in Liverpool, Belfast and Glasgow. As in those areas – and the more prosperous parts of southern England – unemployment now is a much worse problem than it was up to 1979. But (as also in those areas) *relatively* high unemployment has been a continuing characteristic of Thanet over several decades, and cannot simply be attributed either to the recent collapse of manufacturing employment or to the earlier loss of its tourist base. In the 1960s, when absolute levels of unemployment were much lower everywhere, Thanet's problem was seen from outside as essentially one of winter unemployment, less serious in character than the year-round unemployment of declining industrial areas, and suspected to include many prematurely retired men who were not really seeking work. Over the years, however, the seasonal element has shrunk markedly, as the importance of tourism in local employment has declined, and winter unemployment is now only about 10 per cent higher than the summer level – as compared with 20–25 per cent in the majority of seaside resorts and 40 per cent or so in the most specialized resort areas. The main concentrations of unemployment are now clearly among the young, and premature retirement can only account for a very small part of the problem.

Significantly, the level of unemployment in Thanet has remained relatively high even when, as in the period between 1960 and 1966, employment trends in the local labour market have been distinctly favourable, with the effect of manufacturing job creation being offset by substantial in-migration. In 1966, for example, the Census recorded an unemployment rate of 6.4 per cent in Thanet against an average of 2.9 per cent for the country as a whole. The local unemployment rate approached the national level most closely between the mid-1970s and the early 1980s but even then the Thanet rate was about a quarter higher. The worsening of Thanet's position since 1983 clearly reflects the substantial redundancies in local manufacturing and transport. Over the long run, however, there is strikingly little relationship between trends in employment and unemployment, indicating that local unemployment is not simply a question of demand deficiency, but reflects other characteristics of the local labour markets.

In fact, the Thanet labour market seems to have three sets of distinctive characteristics. These derive respectively from its era as a predominantly tourist economy, the weakening of demands for labour during the 30 years' decline in that economy, and the area's continuing attraction to working-aged as well as elderly in-migrants.

On the first point, Thanet's tourist industry has always been highly seasonal and dominated by small establishments, with secure jobs and prospects of career advancement being largely restricted to the self-employed and family workers. More secure employment was principally to be found in population-serving activities with only limited opportunities for careers in professional activities or banking. Employment instability was thus a *normal* experience for workers in the area (men as well as women) and school-leavers seeking skilled employment would leave the area to obtain apprenticeships. This characteristic has persisted through the period of tourist decline and manufacturing growth. In our survey of work histories, we found the median length of a spell of employment in manufacturing to be around a year, with an annual turnover rate of about 40 per cent, which is much above the norm elsewhere. The connection with tourist industry practices is indirect since less than a quarter of those who had worked in manufacturing had also worked in the holiday industry. Almost half the 'unstable' manufacturing workers (those with no job of more than two years duration) had actually come as young migrants to Thanet (under the age of 30) – the remainder being born and bred there. The high rate of turnover appears to affect all the main manufacturing industries in Thanet and to have been a characteristic of the area's manufacturing since at least the late 1960s, rather than being an adaptation to recent competitive pressures. It is thus essentially a structural feature of the local economy, created by its past forms of economic development but maintained now partly by selective migration and partly by the mutually reinforcing expectations of employers and workers. Within Thanet, however, there appear to be some significant variations between residential localities, with a notable difference between the two council estates in our survey, suggesting parallel processes of segmentation in the housing and labour markets.

A related feature of the area is the low level of unionization, which was characteristic of the small establishments and high turnover jobs in the tourist economy, but which persisted through the period of industrial growth. On the evidence of our survey this lack of organization has less to do with rooted objections to union membership than to there being no pre-existing union pressure at the place of work. As Gallie (1987) argues, a lack of union organization in an area in previous generations tends to reproduce itself, whatever the change in circumstances, both because of a lack of early socialization in the tradition and the lack of a base for recruitment.

A second set of effects on the local labour market is due to employment decline. This is primarily to be seen in high levels of unemployment. However the causes of this do not lie entirely in Thanet's own employment fluctuations. For the most part they have to be seen as the result of national, or at least regional, job losses which reduce the possibility of out-migration coupled with a local labour force which is in a relatively weak competitive position. In the current climate we should expect not only the evident limitations of the skill stock among local manual and service workers, but also their typically unstable work histories to count against them in competing for scarce employment opportunities. In this respect, a significant finding of the MSC's Restart interviews with those unemployed for 6 months or more has been the high proportion of the Thanet respondents (69 per cent) lacking any identifiable qualifications, since this group have roughly double the chances of unemployment.

In the early part of the current recession, before 1981, Thanet did not fare too badly. The effects of job losses since then, however, particularly in manufacturing, were clearly visible in the responses to our interview survey. Twenty respondents had worked in manufacturing jobs in 1979 and/or 1980: of these, five were still with the same firm in late 1986, two others were self-employed or family workers in manufacturing, three were now construction workers, four were unemployed (two of them continuously since 1982) and six were at home looking after families. Overall, the proportion of long-term unemployed is not as high as might be expected (38 per cent in October 1986) but would probably be substantially higher were it not for short-term flows off the register to fill temporary jobs during the seasonal peak. What is significant in relation to the experience of our respondents is that few ex-manufacturing workers are any longer in regular jobs in that sector.

The third and crucial influence on the Thanet labour market has been the pattern of migration into and out of the area. Despite low earnings, high unemployment and lack of employment growth, there has been a continuing net *inflow* of economically active (½ per cent of the economically active population in 1980–1), as well as retired, migrants for environmental or personal reasons, which has kept the rate of unemployment higher than it would otherwise be. This balance is the outcome of several distinct flows and counter flows, affecting different age groups. Among young people, particularly in the 15–24 age range, there has long been a heavy net outflow of the more amibitious elements in the local population in search of higher education, training or steady jobs elsewhere in the south-east. Over this age range net out-migration has served to remove about one in six of each cohort entering the working age population of Thanet, but this is made up of a gross loss of one in three counterbalanced by inflows. In the current recession this out-migration appears to have been somewhat stemmed, or at least deferred, by the lack of opportunities elsewhere, for example in the industrial centres to which Thanet youngsters traditionally went for apprenticeships. There has also been a new inflow of uncertain proportions (fluctuating with changes in law and administrative practice) of young unemployed people seeking affordable accommodation within the limits of their lodging allowances. More traditionally the inflow of workers in their teens and twenties has been (to judge by our survey) of people with very similar characteristics to those of their contemporaries who remain in Thanet. After the age of 30, however, the numerical balance of migration becomes more favourable and there is a substantial net inflow of people in their thirties. In this case it is the inflow rather than the outflow which is notably selective in its characteristics: to judge, again, by our survey, the incomers in this age group (including some return migrants) are more petty bourgeois in their orientations, mostly looking for work or businesses in the service sector, and less exposed to unemployment than those who have remained in Thanet. Finally, in the age range between 45 and retirement, there is a further large inflow, comprising essentially pre-retirement migrants: fewer of this group were to be found in the low and middle income areas on which our survey focused, but that is in itself evidence that they are likely to be drawn from the same range of (not particularly senior) non-manual and skilled manual occupations as Thanet's retired population. The ironical outcome of this mixture of inflows and outflows seems to be that the mature in-migrants drawn to Thanet, in many cases by extra-economic considerations, effectively keep the unemployment rate up, but themselves enjoy a relatively strong competitive position which concentrates unemployment to a greater extent among those who were born there or moved there at an earlier age.

Development policy and local politics in Thanet

Our account of economic change in Thanet has already touched on many of the key influences on development policy and local politics in Thanet. In this section, after summarizing these influences, we outline some of Thanet Council's economic development policies and their political repercussions.

The context of development policy has four elements: the prevailing conservatism of the area, the fragmentation of interests among localities and social groups, the divergent attitudes regarding municipal intervention, and the area's economic weakness.

The predominance of private sector employment (especially in small firms), of self-employment, and of owner occupier retirement migrants has made the area a primarily conservative one. Thanet has elected one or (since 1974) two Conservative MPs ever since the war, and Thanet District Council was controlled by the Conservatives from its creation in 1974 to October 1984, since when no party has had an overall majority. The lack of any tradition of large firms, major public sector employers, or unionized employees has meant that Labour support has been weak. Opposition is thus expressed by Alliance and (in local elections) Independent voting as often as by Labour voting.

Secondly, the fragmentation of the area due to its origin as three resort towns has hampered its development, and led to a divisive style of politics. The historical distinctiveness of the three resort towns has been only partly modified by subsequent development. The central location of manufacturing has prevented any one of the towns from specializing in this. But the expansion of Ramsgate port for freight traffic and the passenger/car ferry has given Ramsgate politics a specific stamp. It has reinforced an earlier tendency for the area to be more 'industrial' and to have stronger Labour political activity. The port expansion has created jobs but has also led to a large increase in the volume of lorry traffic through west Ramsgate which contains some of the more 'environmentally sensitive' areas and voters. As we shall see this led to a key conflict within the Conservative Party. Retirement migration has been greatest in Birchington and Westgate which are the centres of Independent support. Finally Margate and Broadstairs are the major centres of Conservative support.

A third influence is the pattern of attitudes to municipal intervention which is closely related to the increasingly diverse social and economic structure. In the heyday of tourism in Thanet, entrepreneurialism and council intervention were combined in the form of municipal Conservatism. There was widespread support for local council interventionism since it produced results. The success of the tourist industry meant intervention was popular among the public and had spin-offs for businesses generally as well as helping the component firms of the tourist economic base. The manufacturing phase, however, brought in firms which did not require a continuing provision of infrastructure and promotion of place of the sort associated with tourism. Likewise, the decline of tourism meant that small firms no longer saw their interests as linked to those in tourism. The expansion of Ramsgate port, strongly promoted by the council, led to further divisions due to the level of spending involved, and the environmental impact of the lorry and other traffic it caused. The environmental effects in particular led to a split among Conservative voters in Ramsgate, many of whom lived in the attractive western districts of the town. Finally, the increasing proportion of retired people has meant the emergence of a population segment with interests tilted away from development and interventionism and towards conservation and low rates.

The result of these economic and social changes is that today, in place of a single successful industry backed up by council intervention and public support, there is economic and social diversity and a decline (though not disappearance) of public support for economic intervention. The local economy and local polity now contain major segments which are opposed to substantial council economic intervention of the sort associated with tourism.

Recent years have thus witnessed an ironic development: there is agreement on the need for council intervention in the economy, but a crisis of legitimacy for any such intervention since there is no longer a consensus about the form it should take.

Finally, council economic policy and local politics are greatly influenced by the weak competitive situation in which Thanet finds itself. As we saw earlier this becomes more significant in a period when 'negative assets' of the kind Thanet has exploited in the past (low wages, available labour, cheap land and housing, and obsolete buildings), are widely available and there is a renewed search for locational advantages. Its effect is to place the council in a weak bargaining position *vis-à-vis* private firms. The generous deals it has entered have often led to accusations of dishonesty or corruption, and a deterioration in the council's reputation. The controlling party is held responsible for these deals, but the fact that councils of different political colours have engaged in the same practices suggests that an underlying factor, economic weakness, is primarily responsible.

Council economic policy

We shall outline policy in two areas: tourism and industry. These will enable us to bring out the connection between restructuring in Thanet, policy options and their political effects.

TOURISM

In the field of tourism, the council engages in resort promotion, encouragement of investment in new hotels and upgrading of existing hotels, municipal entertainment provision, and planning policy in support of tourism. For reasons of space we shall discuss only the last two of these.

Entertainments are a traditional council responsibility in seaside resorts. Thanet inherited entertainment facilities in all three of its towns, but the Winter Gardens at Margate is the largest of them. The council's commitment to tourism is reflected in its policy of keeping open the Winter Gardens despite its regular six-figure annual losses, and in its provision of a 'summer season' of entertainments there for tourists. The main forces keeping the Winter Gardens open are the tourist hotels lobby (and in particular the Margate hoteliers organization) and the council's own commitment. In addition voluntary organizations such as amateur dramatics societies are influential; demand from companies for functions such as dances has disappeared as social clubs have been rationalized out of existence.

The management of the Winter Gardens has been a continuing headache for the council. From the council's point of view the Winter Gardens building is a millstone. Well-adapted to its purpose when built in 1911, it is now ill-suited to demand. Building an alternative with council money would involve a high and risky commitment given the state of the tourist industry; and so far private or joint venture alternatives have not been forthcoming. The level of annual losses is held to reflect on the competence of the entertainments staff employed by the council, the Leisure Committee of the council and the council as a whole.

The main change made in response to the losses has been in the management of the Winter Gardens. At one time entertainments promoters were engaged to arrange the summer season. However by 1981 no takers could be found, and the council took on the organization of entertainments itself. At first the arrangement guaranteed a sum to the artist whatever the actual takings. It was while this system was in operation that the 1985 'crisis' occurred (£150,000 loss). This led to demands for the resignation of the council officer responsible (successful) and of the Leisure Committee (unsuccesful). The council responded to the political rumpus by introducing a risk-sharing arrangement under which council and artist received agreed shares of the takings, e.g. 25 and 75 per cent respectively. This appears to have been successful in keeping losses down in 1986 and 1987.

The Winter Gardens is thus a good example of the way Thanet's heritage affects present day economic and political realities. It imposes a large cost on the council budget and generates a high level of political conflict now that the tourist industry is no longer successful.

The second form of council intervention in support of tourism is *planning policy*. This is best seen in its designation of part of Cliftonville (Margate) as a holiday zone. The aim of this policy is to support tourist hotels, and prevent them converting to non-tourist uses. For hoteliers this 'secures' the immediate environment by freeing it of unemployed people, the mentally handicapped, retirees in nursing homes, etc. It means that they do not have to worry about the 'nuisance' effect of late-night functions. This policy has been popular with hoteliers: their main complaint is that it is only partially effective. In particular, appeals to the minister against council refusals of permission to convert hotels have often been successful, and the widening of the General Development Order under which certain types of development do not require planning permission reduces council control. The main opposition to the policy comes from people wanting to convert from tourist to non-tourist accommodation (residential homes, for example). The policy is alleged to enforce bankruptcy by denying tourist hoteliers an alternative livelihood. For the planning department it is an important tool and symbol of the potential of planning powers. In May 1986 a more flexible version of the policy was agreed (and later approved by Kent County Council since it had been embodied in the Local Plan) to try and reconcile the conflicting interests.

This policy shows vividly the problems of managing restructuring as the new and old uses of hotels vie with one another for survival. The policy has not however led to wider political controversy as has entertainments policy.

INDUSTRIAL POLICY

Thanet Council's industrial policy includes a steady programme of building small industrial units, a loans fund for aiding industry, and the provision of rate-free periods to business. It has used planning policy to deny permission for non-industrial development on former industrial sites, though often such decisions have been overturned on appeal. But net spending on industrial estates, industrial promotion and aid to industry is a small part of the council's budget – £100,000 in 1985/6. This compares with net tourism-related spending of £987,000

Plate 5.4   Ramsgate port, Sally Line ferry.

on theatres and foreshore activities, and £113,000 on resort promotion (not to mention parks and sports activities).

The main industrial initiative, however, has been *port-related*: the
support given to Sally Line cross–Channel ferry, and the subsequent
Sally Line management of 'Port Ramsgate'. This is a long story but
worth outlining since it reveals a great deal about the politics of
economic development policy in Thanet.

The essential facts are these. Ramsgate was a pleasure port with a
marina and small commercial port until the mid-1970s. A plan was
then launched to expand it by reclaiming land and building a sea wall
to enclose it. This led to a ferry service run by the Norwegian Olau
Line from 1979 to 1981 and by Finnish–owned Sally Line since 1981.
Sally Line's ability to compete with the Dover services is based firstly
on carving out a distinctive section of the market. The sailing time is 3
hours (compared with 1¾ hours) but the price is generally lower
(particularly for a car plus 4 passengers). In the freight market Sally
attracts hauliers who leave their loads for transhipment, with haulage
from Dunkirk done by another vehicle – Ramsgate's advantage over
Dover here is its space for parking loads before and after crossing. It
has also been very innovative in developing ways of running ferries in
the off-season.

The ferry service has generated jobs and this accounts for both its support among corresponding groups of the population, and the council's determination to support it. The number of jobs according to Sally figures is 500, though the number of direct employees is less than half that. But it has also generated large capital costs and is an environmental nuisance: these factors account for its unpopularity among some ratepayers and among residents in areas directly affected by the flow of lorry and other ferry traffic.

By March 1986 the council had spent £7¼ million on the sea wall and land reclamation, and on dredging channels for the Olau and Sally Lines. This sum includes £1¼ million in the form of capitalized interest consequent on the closure of the Olau Line and loss of revenues from it. Revenues from the project are insufficient to cover the interest charges and capital repayments, with the result that between 1982/3 and 1985/6 annual losses ranged between £263,000 and £647,000 (half of the latter figure was premature debt redemption). They have since fallen. These amounts are much larger than those expended on aid to industry, but much smaller than those spent in support of tourism.

The council is thus seen as having committed large sums of public money to a project which does not cover its costs from the council's point of view. Controversy has focused on the level of the cost and on the agreement with Sally which is widely believed to be too generous (Sally has an exclusive right to run ferries from Ramsgate, and since becoming port manager has been able to control the port environment to its advantage). Subsequent expansion plans have also been vigorously opposed, and dominated Thanet politics in 1985 and 1986: they consisted of a plan to reclaim more land to expand the size of Port Ramsgate, a new access road to the port, and in June 1986, a plan for a rail link to the ferry terminal. The expansion plan and access road were supported by the council and by those who might benefit from the extra jobs. There was also support for a better access from those dissatisfied with the proximity of large lorries and hazardous loads – and opposition to any increase in the volume of lorry traffic. The rail link plan also divided the population. In this case the Planning Department of the council was also opposed, considering that environmental detriment outweighed job gains. Residents in Cliffsend, where a rail link would have been built, were strongly opposed and created or reactivated residents' protest groups.

In the event, the rail link was abandoned in December 1986 after BR and SNCF chose Dover over Ramsgate, but the style in which Sally pushed for a fast decision, and the council went to the legal limit to meet Sally's demands, confirmed opponents' suspicions regarding the council. Sally's conduct regarding the access road was to withhold signature from an agreement on sharing costs (Kent County Council £1m, Thanet District Council £350,000, Sally £350,000 is the likely breakdown) pending receipt of planning permission. Once this was received, a new condition was added – it wanted a coastal protection grant before signing. The result was that for a period of some months the access road had received the go ahead for 1987/88 (a remarkably fast ascent up the scale of priorities due to concerted lobbying and a 'conscience' by Kent County Council about its past treatment of Thanet) while Sally had not signed any agreement. It has now done so. At the time of writing the results of the June 1987 public inquiry into this access road have not been announced.

In sum, having made the initial commitment to the Ramsgate port expansion the council has been in a weak bargaining position *vis-à-vis* Sally. Its financial wisdom has been questioned, as have the job projections of the latest plans. The fact that in February 1986, when the access road/expansion plan was being hotly debated, councillors accepted a free cruise from Sally in celebration of its fifth year of operation, with presents for all concerned, confirmed opponents of the Thanet District Council/Sally link in their feelings. It is the most striking illustration of the sequence we identified earlier: a weak structural position, generous deals with private enterprise, and a deterioration of the council's reputation.

Political repercussions of council policy

Thanet's tourism and industrial policy has thus centred on two costly items – support for entertainments provision and subsidization of the Ramsgate port expansion.

The political effects of these policies have been twofold. There has been a deterioration of the council's reputation due to the accusations of mismanagement (in the case of the Winter Gardens) and dishonesty and corruption (over Ramsgate port), though there is a lot of support for council economic intervention in general. Secondly, economic policy has exacerbated the conflict between supporters and opponents of intervention in the Conservative Party, and it was this which led to the demise of Conservative control in 1984. In brief, the dominant group in the Conservative pressure on Thanet council from 1974 onwards, known as the 'Old Guard', was Ramsgate-based and interventionist – it was responsible for the port expansion. The leaders of the anti-intervention group (the 'rebels') were from Broadstairs and Ramsgate but the bulk of its support came from Margate. A series of scandals involving 'Old Guard' councillors occurred in the 1977–83 period, some of which made the national headlines. These included the placing of contracts with firms in which councillors had an interest, counterfeiting money, prostitution, and the perpetration of a ruse in which an oil sheikh appeared in Ramsgate with a view to putting forward a rival bid for the port at a time when the Sally Line was engaging in brinkmanship before signing a contract with the council. As a result of these scandals the Old Guard lost support to the rebels who eventually formed a majority of the ruling Conservative group after the May 1984 elections.

Once in power the rebels set about implementing their policies of privatization of council services, sales of assets, low rates and a low level of council intervention. But in Ooctober 1984, faced with the blunt choice between selling assets or sacking staff, the Tory group voted against the controlling rebels and deposed the leader. The two Thanet Conservative Associations retaliated by securing the expulsion of those involved in the coup (who were now referred to as Unofficial Conservatives). The result was that from October 1984 the Conservative group was split into equal numbers of Official and Unofficial Conservatives (14 of each) and hence lost its overall control of the 54-seat council. Since that time no party has had an overall majority.

Development-related policy disputes are thus central to local politics. There is considerable public support for council economic intervention in the abstract. However, specific efforts arouse controversy among different groups on grounds such as mismanagement, corruption, cost or environmental effects. The fact of the split among Conservatives over interventionism symbolizes the political significance of the changes in the local economy since the heyday of tourism and municipal conservatism.

Conclusion

What does our account of Thanet's economic development, local policy and politics tell us about the problems of localities and their chances of solution?

A key factor about Thanet, which in itself constitutes a 'problem', is that there is no agreement about what precisely its problem is. One interpretation is that its unemployment level is due to an inability to attract new investment to replace employment in industries in decline (tourism, manufacturing). This emphasizes the changed context in which Thanet is competing. Its previous assets (cheap and abundant labour, cheap housing) are no longer so valuable now that more areas possess them and the manufacturing sector of the economy is in any case in decline. Likewise in the services sector the Thanet towns cannot hope to share in the growth in financial services employment since they have so few administrative control functions and are low in the urban hierarchy. The strategy of creating or exploiting locality-specific advantages follows from this analysis and Thanet council has done this via its support for port expansion, the development of Manston airport, etc. It has also pressed for regional assistance (unsuccessfully) and improvements to the Thanet Way road link to the M2 to London (successfully).

A second interpretation is that the level of unemployment would remain high even if new investment was attracted since the jobs would be taken by migrants rather than locals. This is consistent with the evidence presented earlier on the disjunction between unemployment level and employment growth. The argument here is that the local population is unlikely to take the likely jobs and that this is due to a combination of factors: (a) the lack of skills due to the deskilling effect of the dominance of 'secondary labour market jobs' (such as hotel or nursing home work); (b) the effect of cheap housing in attracting unemployed people to Thanet from a wider area; and (c) the low wage levels which mean that the incentive to work for those (especially adults with children) on supplementary benefit is low. This interpretation stresses the concentration of social disadvantage in Thanet. In a sense it is arguing that in part Thanet's problems are 'imported' from other areas. It has formed part of the council's case for regional assistance, but no specific measures have been addressed to it.

The question of the nature of Thanet's problem leads naturally to the question of who takes action over it. What is lacking in Thanet is a broad-based 'spatial coalition' in which employers, unions and council work together to promote the prospects of the area (Pickvance, 1985). Instead it is the council which has taken the lead on its own. A key factor in explaining this weak level of area-wide representation is the absence of big capital. Ironically, in areas with larger firms the declaration of large-scale redundancies by these firms can be the trigger for a widely based mobilization in which the big firms may even take a leading part. In Thanet, on the other hand, the small and medium size of the firms has prevented this happening. The lack of large firms and, especially, large public employers also contributes to the lack of a broad-based spatial coalition. The absence of high levels of unionization characteristic of these types of establishment means that trade unions which might take an active part in spatial coalitions are missing. Conversely the small scale of most business contributes to the fissile nature of local politics. Again and again we have been struck by the inability of 'Thanet' to act collectively. Until recently its representatives on Kent County Council have been ineffective, and frequently it has not pressed for benefits potentially available to it. In brief, even if there was agreement on the problems facing Thanet, the lack of organizational capacity would limit the chances of addressing them.

Finally the question of the economic preconditions of future prosperity in Thanet needs to be posed. Whatever its peculiarities, such as the impact of the tourist legacy on labour markets and politics, Thanet falls within a much wider category: that of places which lack or have lost any special role within the urban and regional system, and through this lack the assets (capital and organizational capacity) to restructure themselves. Such places are particularly dependent on and vulnerable to the actions of the external capital and national government. The role of local government in this depends on two conditions. It requires public support which may take the form of a political consensus or a spatial coalition – neither of which exists in Thanet – and it needs to 'lever in' investment from external sources.

Thanet's current problems in the 1980s are not simply the result of local decline but of national recession and high unemployment. The main benefit of faster national economic growth would be that Thanet imported less unemployment. This would be a substantial benefit. More positive paths to economic restructuring for places such as Thanet face considerable difficulty. A market-led strategy which in the case of Thanet would include a further growth of subsidized provision of welfare services might provide limited extra employment but would do little to resolve the problems of the labour market or of political capacity. If such places are to move away from dependence on cheapness as their only asset, and find some stronger economic base, then a leading role for the public sector in defining a strategy and giving it support is essential. This chapter has suggested how the history of one such place leads to the build-up of political as well as economic impediments to restructuring: indeed, an economy based on small businesses, private services and disorganized labour seems to be incapable of restructuring itself. This may be a disturbing conclusion when the nation is being encouraged to move in the same direction.

References

Canterbury and Thanet District Health Authority (1985) *Strategic Plan: 1984/5–1993/4*.

Canterbury and Thanet Community Health Council (1986) *Canterbury and Thanet Residential Homes Guide*.

Challis, D. (1984) 'The Evaluation of Social Work', Ph.D. Thesis, University of Kent at Canterbury.

Gallie, D. (1987) 'Patterns of similarity and diversity in British urban labour markets: trade union allegiance and decline', paper to the Sixth Urban Change and Conflict Conference, University of Kent at Canterbury, September.

Judge, K. (1985) 'Caring for profit', Personal Social Services Research Unit Discussion Paper No. 311, University of Kent at Canterbury.

Kent County Council (1952), *Kent Development Plan (Part A): Report upon the survey and analysis of the problem, Volume 1.*

Martin, R. T. and Ellis, M. F. (1982) '*The Thanet dimension*', paper presented at the Divisional Conference of Head Teachers, Eversley, Kent.

Pahl, R. (1985) *Divisions of Labour*, Oxford: Blackwell.

Pickvance, C. G. (1985) 'Spatial policy as territorial politics: the role of spatial coalitions in the articulation of "spatial" interests and in the demand for spatial policy', in G. Rees et al. (eds), *Political Action and Social Identity*, London: Macmillan.

Pimlott, J. A. R. (1947) *The Englishman's Holiday*, London: Macmillan.

Stafford, F. and Yates, N. (1985) *The Later Kentish Seaside*, Kentish Sources IX (1840–1974), Gloucester: Alan Sutton for Kent Archives Office.

Urry, J. (1988) 'Cultural change and contemporary holiday-making', *Theory, Culture and Society*, Vol. 5, pp. 35–55.

6

Paradise Postponed: the Growth and Decline of Merseyside's Outer Estates

RICHARD MEEGAN

The social and economic spatial restructuring of Merseyside

Introduction

Published in the final year of the Second World War, the *Merseyside Plan 1944* was clear on the task facing the postwar planning of the area:

> The decentralisation and regrouping of the population displaced on the reconstruction of the congested areas of Central Merseyside, in conjunction with the distribution and location of the new industrial areas constitute the main regional planning problem. (Thompson, 1945, p. 4)

The economic future of Merseyside was thus seen to lie in the decentralization of industry and the matching dispersal of population. The idea was not new. Liverpool City Council, a decade earlier, had come to a similar conclusion and had sought from Parliament special powers to restructure its industry and population to this end. The Liverpool Corporation Act (1936) enabled the local authority to buy up land on its outskirts and to build on it factories for sale or rent and offer incentives to industrialists to do likewise. Two new industrial estates were begun at Speke, to the east of Garston dock, and at Aintree, to the north-east of the city centre, and negotiations begun for a third, to the east of Aintree at Kirkby. Sites for housing were also acquired in Cantril Farm, Huyton and Speke.

While these municipal ventures were effectively interrupted by the onset of war, the preparation for and duration of the conflict ensured that some development did take place in these newly industrial areas with the location by central government of important armaments and war-related production in them. Thus, Speke was the site of an important 'shadow' airframe factory run by Rootes while the then rural area of Kirkby was completely transformed by the erection of a massive shell-filling Royal Ordnance Factory. And postwar reconstruction on the lines of those proposed by the *Merseyside Plan, 1944* was to ensure that the importance of the outlying areas as production centres was to be sustained by the continuing dispersal of jobs and people to them.

The changing geography of production: the 'growth years' of the 1950s and 1960s

Conversion to civilian production after the war proceeded rapidly. The airframe factory at Speke was allocated to Dunlop to manufacture rubber tyres and footwear while the Kirkby Royal Ordnance Factory was acquired by Liverpool's corporation to enable it to fulfill its original plan to develop an industrial estate in the area. Subsequent growth was initially based on firms relocating from inner Liverpool, although many of these moves proved to be relatively short-lived with a substantial number of the small firms that dominated this first postwar wave of investment either closing or moving on (Gentleman, 1970; Lloyd, 1970). Gradually however, national and multinational companies, many new to the area, began to set up. One – Albright and Wilson, the chemical manufacturer – even went so far as to bring its own premises with it, dismantling a wartime flying boat factory located at Windermere and transporting it to the Kirkby Industrial Estate for re-erection! Other newcomers to the Kirkby industrial estate in the years immediately following the war included Kodak, BICC and John Dickinson (manufacturing chemicals, electrical cables and stationery respectively). But the companies in this first wave of investment did not employ anything approaching the numbers previously working on the Royal Ordnance site. At the end of 1950 they employed around 4700 workers, only a fifth of the wartime workforce. The Dunlop factory in Speke alone employed more than this (over 7000 workers).

The situation was to change, however, in the 1950s, with an influx of companies taking advantage both of the area's 'Development Area' status – encouraged in this context mainly by the use of Industrial Development Certificates to control growth in the 'overheating' south and Midlands (Lister, 1983) – and of the labour force finally being dispersed to the area by the city's corporation (Roberts, n.d.). The big national and multinational corporations that made up this second major wave of investment in the Kirkby industrial estate included Birds Eye, A.C. Spark Plug (now Delco Electronics, a subsidiary of General Motors), Fisher Bendix, Kraft, Otis Elevator (a subsidiary of United Technologies) and Yorkshire Imperial Metals.

By the end of the decade these and other firms in the estate were employing nearly 16,000 workers. Further south, in Speke, around 13,000 were employed in local factories. A particularly important sector was chemical manufacture which grew rapidly in the 1950s, encouraged not least by the establishment of the National Health Service and building in some cases on wartime manufacture of drugs like penicillin. Major employers included Distillers (established in 1952 and taken over by Dista in the early 1960s) and Evans Medical (set up in 1944 and subsequently taken over by Glaxo).

It was the rapidly expanding car industry which provided the next major infusion of manufacturing investment in the outer areas in the early 1960s. Again a combination of regional policy and labour availability was important in the expansion of the British Motor Company's components factory at Kirkby and the location of car assembly plants in Speke (Standard Triumph), in the still relatively undeveloped Halewood area adjacent to Speke (Ford) and in Ellesmere Port (Vauxhall). This wave of motor manufacturing investment rounded off dramatically the transformation of the outer areas set in train by the war; bringing with it the promise of more jobs – over 30,000 – than were then provided by both the Speke and Kirkby industrial estates added together.

In just two decades the Merseyside economy had been significantly restructured both sectorally and spatially. While the rundown of the investments in the port and port-related industries and services based on Empire and the 'New World' continued apace in the 'core', investments in the postwar growth sectors in new consumer and capital goods production were being positioned on the 'periphery'. The locus of economic growth and dynamism shifted accordingly. The bulk of the new jobs was to be found precisely where the pre-war and wartime politicians and planners had wanted them to be – on the city's periphery. So, increasingly, were the people.

The dispersal of the 'overspill'

The *Merseyside Plan, 1944* estimated that over 258,000 people in Central Merseyside (about one fifth of the total population) made up what it termed 'overspill' – '. . . the number of persons who cannot be reaccommodated within the limits of the existing built-up areas' (Thompson, 1945, p. 17). Of these, 148,000 were Liverpudlians. So the 'overspill' was dispersed – in different stages but mainly in the 1950s and 1960s – to an outer ring of municipally owned housing estates (Figure 6.1) which, after the main phase of development in the late 1960s, housed between 160,000 and 200,000 people, about one in eight of Merseyside's population (Merseyside Council for Voluntary

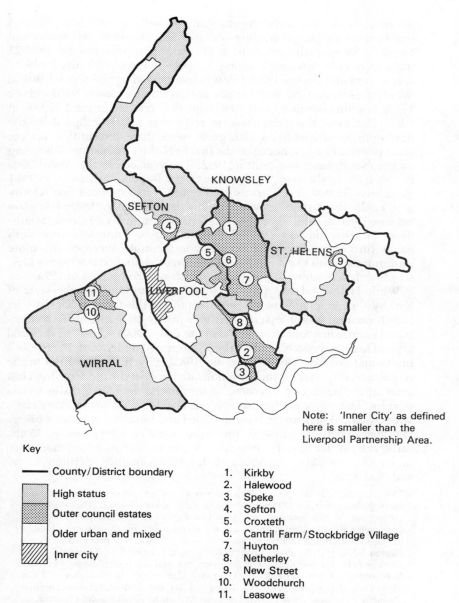

KNOWSLEY

SEFTON

ST. HELENS

LIVERPOOL

WIRRAL

Note: 'Inner City' as defined
here is smaller than the
Liverpool Partnership Area.

Key

—— County/District boundary

High status

Outer council estates

Older urban and mixed

Inner city

1. Kirkby
2. Halewood
3. Speke
4. Sefton
5. Croxteth
6. Cantril Farm/Stockbridge Village
7. Huyton
8. Netherley
9. New Street
10. Woodchurch
11. Leasowe

Figure 6.1   Merseyside's 'social areas'. Source: Merseyside County Council.

Services, 1978). The three largest, – and the ones on which we focused our study – are, in descending order of size, Kirkby, Halewood and Speke.[1] The growth rate of the new estates was phenomenal. In 1921 there were only 366 people living in Speke. Municipal house building was started there in the 1930s but was interrupted by the Second World War. At the end of the War the population stood at nearly 8000 and the 1950s saw this number more than tripled: it reached around 27,000 in 1961. But even this impressive increase was totally eclipsed by the development of Kirkby, which grew more than sixteenfold over the same period. The first house on the first of Kirkby's constituent housing estates (Southdene) was built in 1952. The year before that about 3000 people lived in the area. By the time of the Population Census in 1961 there were 52,000. The Census itself noted in its foreword that Kirkby was easily the fastest growing township nationally. Indeed it soon became clear that the three housing estates originally planned (Southdene, Northwood and Westvale) would not be enough to cope with either further rehousing of people from central Liverpool or, more importantly, with the 'overspill' of the three original estates themselves, so a fourth was added, Tower Hill, which was completed in 1974.

While Halewood was not developed on anything like the scale of Kirkby – its population in 1971 was less than a third that of Kirkby – its transformation was at least qualitatively just as dramatic, being concentrated mainly in a four year period in the early 1960s when '. . . the rural idyll of hedges, broccoli and carrots disappeared underneath Liverpool's bricks and concrete' (Fletcher et al., 1982). Given the emphasis in the population dispersal on those in 'housing need' in the old working class areas of the inner city, the new 'communities' had very distinct class profiles. They were predominantly 'one class' townships: working class, with a marked bias towards semi-skilled and unskilled manual workers (Meegan, 1988). The new communities were also very young. In the early years of Kirkby, for example, 48 per cent of its population were under 15 years old – prompting such local nicknames as 'Bunnytown' and 'Kidsville'.

[1] ESRC support for this project ('The impact of socio-economic change on outer Merseyside', reference D04250013) is gratefully acknowledged. The work was carried out in collaboration with Jane Lewis (formerly of the Department of Geography, Reading University, and now with the Economic Development Unit of the London Borough of Ealing). Research assistance was also provided by Frank Banton (Ockenden Venture, Liverpool), Bernard Charles S.J. (Liverpool), Ann Clark and Heather Smith (Training for Development, Liverpool), Margaret Pearson (Department of General Practice, Liverpool University), Brian Bailey and Ian Thompson (Department of Civic Design, Liverpool University) and Lorraine Donnelly (Liverpool Polytechnic). The usual disclaimers absolving these individuals from responsibility for any errors of fact and interpretation apply of course. The photographs that appear in this chapter were taken by Alex Corina, except for 6.4 which was supplied by Cooperative Development Services, Liverpool.

The dispersal was undeniably traumatic, both for those moved and for those already living there. For the former, long-standing, tightly knit communities were broken up and extended family and kinship networks ruptured as families and neighbours were scattered in the different outlying estates:

'They should have moved whole streets together from Liverpool to Kirkby instead of throwing people together from different streets . . . communities would have been kept intact.' (A 47-year-old woman who moved to Kirkby in the late 1960s. Interview, 1986.)

'It was a very close-knit community and everybody knew everybody's business. I remember around eleven o'clock in the morning – your door was always open and all the women would come round and chat. One would bring a bit of milk, another would bring some bread, sugar, another the tea. They'd come round in slippers – you know, with a bobble on. I remember us having sing-songs . . .' (A 46-year-old woman, living in Speke, recalling her childhood memories of the Dingle area of Liverpool. Interview, 1988.)

'In our house the key was on a string by the letter box. Neighbours could reach in, pull it through the letter box. The priest would just let himself in . . . The street used to meet its own needs in many ways. Each street had a woman who laid out the bodies. There was always someone you could go to who knew how to mend shoes, or do a bit of carpentry, even though they might not be 'qualified'. Old women used to have remedies for sick children . . .' (A 36-year-old man now living in a Halewood council flat reminiscing about his childhood in the 1950s, also in 'the Dingle'. Interview, 1987.)

Countering all this disruption and unsettling change, however, were the hopes of many that the move would mean healthier housing and a chance of more settled and better paid employment, especially for their children:

'It is a new way of life when we come here: we don't benefit but our children will.' (Kirkby resident quoted in Pickett and Boulton, 1974, p. 73).

It is perhaps through the eyes of those children that the sense of adventure, of pioneering a new and better way of life can best be appreciated:

'We came on the waggon train . . . when it was green and lovely,.' (A 41-year-old woman who moved to Kirkby when she was 9 years old. Interview, 1986.)

'It was really strange . . . it felt like we were on our holidays with the green fields and the woods.' (A 38-year-old woman who moved to Kirkby when she was 5 years old. Interview, 1986.)

The original residents in the 'overspill' areas viewed the newcomers with some concern. In Speke it was the families who moved there in the interwar years who had to come to terms with the rapid postwar inflow of younger, larger families. In Kirkby and Halewood it was the rural inhabitants who were faced with the prospect of having to assimilate Liverpool's urban working class. The Chairman of the local Parish Council voiced the fears of the many objectors to the proposed municipal scheme, arguing that it would both '. . . depreciate property and change the whole character of the area' (quoted in the *Liverpool Daily Post*, 29 March 1957). A Halewood man that we interviewed admitted, however, that many of these fears proved groundless:

> 'We expected the scruff of Liverpool but only a minority were "scallies".'
> (A 51-year-old Halewood man, the son of a former local farmer. Interview, 1986.)

The labour market: the 'boom' years of the 1950s and 1960s

While the planners never intended that the outer estates should be totally self-sufficient in terms of jobs, in the early days there was, nevertheless, a more marked mismatch between those seeking work locally and the local supply of jobs than had been expected. The original estimates of the population in the overspill estates proved much too conservative and the timing of the population dispersal did not help either. By the time the first housing was being constructed in Kirkby some of the first wave of manufacturing firms on the industrial estate had been operating for six years and had already built up relatively established workforces.

There was also very little non-manual work in Kirkby, particularly given the fact that there was a five to ten year gap between the first wave of settlers in the area and the completion of the main shopping and leisure facilities. Clerical work in Kirkby in this period was therefore largely confined to the officers of the firms on the industrial estate which were unable to compete with the broader variety of work and more agreeable working environment offered by the service sector in central Liverpool. Consequently, workers were still having to travel back into Liverpool for work and for many, in the early days, '. . . life consisted almost entirely of sleep and work' (Roberts, n.d., p. 16). Some found the strain too great and returned to central Liverpool. Overall, something like 50 per cent of Kirkby's workforce in the first decade of the town's existence worked outside the area, chiefly in Liverpool. This compares with an average for New Towns at that time of about 30 per cent and a figure as low as five per cent for Crawley (Pickett and Boulton, 1974). But as the second wave of manufacturing companies that moved into the area in the late 1950s and early 1960s began to build up its workforces, more residents of the estates began to work locally. Unemployment fell to about five per cent in the mid 1960s and local people experienced what was for many a degree of hitherto unknown prosperity. The change was particularly marked for women workers, many of whom, for the first time, had a wide choice of relatively well-paid and full-time manufacturing work.

But these 'boom years' had produced a very specialized local labour market with two dominant characteristics (Meegan, 1988). First, given the emphasis of the dispersal policy on industry, the jobs were predominantly in manufacturing. In Kirkby, for example, 71 per cent of local jobs were in manufacturing in 1971, just under twice the national figure. The second main feature of the labour market was the extent to which the manufacturing jobs were dominated by British and foreign multinational capital. These two aspects of the local labour market, its dependence on manufacturing and its close integration with the world economy through multinational corporations, made it particularly vulnerable in the economic restructuring that has been prompted by, and is part of, the global economic crises of the period since the mid 1970s.

The restructuring of the 1970s and 1980s: from boom to slump

The 1970s saw the culmination of the longer-term shift of the British economy away from Empire and Commonwealth connections towards Europe. Liverpool's status and role within the global economy shifted accordingly. The most obvious symptom of this changed role was the hastened, albeit long predicted decline of the port as Liverpool, to use Tony Lane's rather apt seafaring metaphor, was left '. . . marooned on the wrong side of the country' (Lane, 1987, p. 46). The city's docks had still been handling about 23 per cent of Britain's manufacturing exports in the late 1960s but this figure was down to less than 9 per cent within the space of a decade. From being the second largest UK port in 1966, Liverpool fell to sixth place in 1986 behind the now more favourably sited ports on the eastern coastline (Lane, 1987) Employment in dock work and sea transport, already falling in the 1960s, fell by a further three quarters between 1971 and the mid-1980s, when almost 20,000 jobs were lost. But the new manufacturing sectors, which the planners had hoped would compensate for the decline of the port, did not stick to the script that they had been given. Within the overall shifts in the world economy, manufacturing was in crisis – and particularly UK manu-

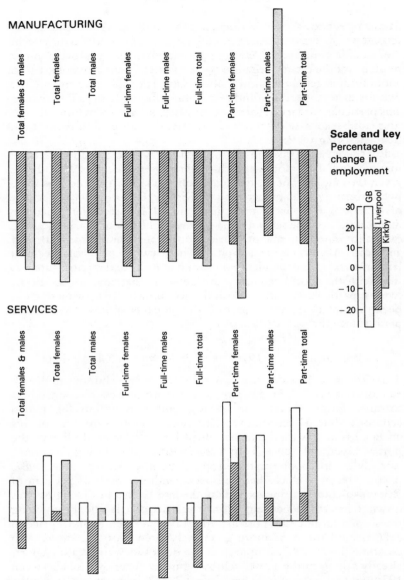

Figure 6.2 Percentage employment change in manufacturing and services: Great Britain, Liverpool and Kirkby by male and female, full-time and part-time, 1971–1984. Source: Annual Census of Employment. Calculated on the basis of the percentage changes between 1971 and 1981 (1968 SIC) and between 1981 and 1984 (1980 SIC).

facturing. Merseyside did not escape the restructuring that followed. Figure 6.2 compares manufacturing employment change between 1971 and 1984 in Great Britain, the Liverpool travel-to-work area and, within the latter, Kirkby. Between these years, manufacturing employment fell nationally by just under a third. In Liverpool as a whole, it was more than halved as something like 95,000 jobs disappeared. Given their relatively heavy dependence on manufacturing, this loss bore particularly heavily on the overall areas of the 1950s and 1960s. As Figure 6.2 shows, in all of the categories of change shown but one, the loss of jobs in Kirkby was proportionately greater than both the national decline and that for Liverpool as a whole. The single exception is itself particularly revealing. Although the actual numbers were relatively small (less than a hundred jobs) and dwarfed by the overall decline (of some 13,000 jobs), it is notable that the sole category of growth, and divergence from the national pattern, was in part-time work, with the obvious implications for levels of pay in what were already relatively low-income communities. The overall loss of 13,000 manufacturing jobs represented a 57 per cent decline in Kirkby's manufacturing employment base. The promise of secure, well-paid jobs that the new manufacturing companies had seemed to offer in the 1950s and 1960s was dashed – especially for women. In Kirkby, the female workforce in manufacturing was slashed by 62 per cent (compared with a national decline of 33 per cent). Men's employment fell by 53 per cent (compared with a 32 per cent fall in Great Britain as a whole).

As in the other localities discussed in this book, the production reorganization which underlay these employment changes was a complex interplay between the macro-economic pressures at national and international level and factors unique to the locality itself. A number of features of this restructuring stand out. There was, first, the widespread scrapping of 'obsolete' capacity. A marked feature of Merseyside manufacturing in general was the relatively low level of investment in new capacity over the 1970s (CES Limited, 1985). Thus in many key sectors the major waves of investment in the 1960s had, for want of replenishment, been left to work themselves out. This showed itself in the collapse of small- and medium-sized companies but in terms of jobs lost, the scrapping of outworn capacity had its most dramatic effect in the large branch plants of multinational companies as they juggled their production across the country and across the world.

Table 6.1 gives some idea of the impact of the production reorganization and the role of the multinationals in it. It shows the major job losses in Kirkby, Speke and Halewood in the period since 1978, when the restructuring was at its height. Some 20,500 redundancies are listed, with more than 90 per cent of these being accounted for by multinational corporations.

Table 6.1 *Major redundancies (150+) and plant closures in Kirkby and Speke, 1978–1987*

Year	Company (Ownership)	*=MNC	Location	No.	Comments
Food, drink and tobacco					
1978	Birds Eye (UK, Unilever)	*	Kirkby	450	
1981	Cousins Bakery (UK)		Speke	260	Closure
1981	Kraft (US)	*	Kirkby	370	
1982	Kraft (US)	*	Kirkby	930	
1983	Seagram (Canada)	*	Speke	220	
Chemicals and allied industries					
1979	Evans Medical (UK, Glaxo)	*	Speke	230	
1980	AKZO Chemie (Holland)	*	Kirkby	115	
1984	Synthetic Resins (UK, Scott Bader Commonwealth)		Speke	125	Closure
1986/7	Glaxo	*	Speke	450	
Metal manufacture					
1981	Yorkshire Imperial (UK, IMI)	*	Kirkby	200	
1981	Yorkshire Imperial (UK, IMI)	*	Kirkby	317	
Mechanical engineering					
1979	KME (UK, Silverines Ltd; ex-Fisher Bendix and KME Workers' Cooperative)		Kirkby	700	Closure
1980	Ward and Goldstone (UK)		Kirkby	160	Closure
1981	Otis Elevators (US, United Technologies)	*	Kirkby	125	
1982/3	Cross International (US)	*	Kirkby	355	
Electrical engineering					
1978	Plessey (1) (UK)	*	Kirkby	380	Closure
1978	Plessey (2) (UK)	*	Speke	330	Closure
1979	BICC Connolly (UK)	*	Kirkby	500	Closure
Vehicles					
1978	Triumph (UK, British Leyland)	*	Speke	4600	Closure
1980	AC Delco (US, General Motors)	*	Kirkby	370	
1980	Ford (US)	*	Halewood	400	
1980	Massey Ferguson (Canada)	*	Kirkby	550	Closure
1981	AC Delco (US, General Motors)	*	Kirkby	159	
1980	Ford (US)	*	Halewood	260	
1981	Pressed Steel Fisher (UK, BL)	*	Speke	900	Closure
1983	Ford (US)	*	Halewood	1300	
1983	Ford (US)	*	Halewood	600	
Clothing and footwear					
1978	F.D. Centre (Switzerland, Starlux)		Kirkby	200	Closure
1981	Commonwealth Curtains (Canada)		Kirkby	131	Closure
Timber and furniture					
1978	Hygena (UK)		Kirkby	200	
1980	Hygena (UK)		Kirkby	300	
1982	Hygena (UK)		Kirkby	700	Closure
Paper, printing and publishing					
1980	Metal Box (UK)	*	Speke	300	
1981	John Dickinson (UK, DRG)	*	Kirkby	214	
1983/4	Metal Box (UK)	*	Speke	218	
Other manufacturing					
1979	H.Hunt and Co. (UK)		Speke	150	
1979	Dunlop/1 (UK)	*	Speke	2300	Closure
1980	Dunlop/2 (UK)	*	Speke	233	Closure
1981	United Reclaim (UK, Dunlop)	*	Speke	153	Closure
TOTAL ALL MANUFACTURING				20,455	

* MNC = Multinational Corporation
Note: The list is restricted to job losses and plant closures in the three study estates. It therefore excludes those occurring in other outlying areas (most notably Aintree, Huyton and Prescot) and, of course, any in the inner city.
Source: Merseyside County Council, company interviews, national and local newspapers.

Plate 6.1   Industrial dereliction, Knowsley industrial park.

From amidst all the wreckage of scrapped capacity, company failures and massive job loss, a 'core' of companies emerged, however, in which there had been substantial investment in new production processes (CES Limited, 1985). It is in the companies that make this 'core' that moves towards 'functional flexibility' and production reorganization based on new 'flexible' production technologies are being variously attempted. Major investment programmes have been associated with the introduction of multi-skilling agreements (as, for example, at Dista in Speke), the removal of a whole middle tier of supervisory and ancillary production workers (as at Metal Box in Speke), and the reorganization of production into separate 'focused workshops' acting as individual profit centres (Otis Elevators in Kirkby). In other cases investment has been directed towards the introduction of 'post-Fordist' production processes (for example, the flexible machining systems in, appropriately enough perhaps, Ford's Halewood transmission plant, or the computerized automatic testing equipment in Delco Electronics' shortened assembly lines).

What all this reorganization adds up to for the workforce has been a marked increase in the intensity of work. While considering themselves to be 'lucky' still to be in employment, most of the workers we spoke to felt that the pace of their work had increased in recent years and, while again recognizing that, on local comparisons, their pay was relatively good, many considered that in real terms this had been steadily eroded as annual pay rises had been progressively reduced and, in some cases, frozen. And, of course, many pay negotiations (like Ford's recent two-year agreement) have been directly linked to changed working practices.

> 'The supervisors are on the girls' shoulders all the time . . . the management want quality and quantity, but they can't have both.' (47-year-old woman)

> 'Somebody from the office, from time and motion comes and says 'Oh, but it can be done, it can be done on paper – but half the time it's an impossibility . . . it's all right pushing you on paper, but when it comes down to doing it eight hours a day, it's a different thing altogether.'

> 'They're pushing girls to the end, they're [the 'girls'] just getting more and more pushed on to them.' (53-year-old woman)

> 'I'm working harder but I feel I've gone backwards paywise.' (38-year-old woman) (Interviews, 1986.)

Alongside the reorganization of manufacturing production, private and public services have also undergone a major restructuring. As Figure 6.3 shows, while employment in services grew nationally between 1971 and 1984 (by some 19 per cent), in Liverpool it actually fell (by nearly 13 per cent). The massive job loss in the docks dominated this contraction (accounting for 53 per cent of the net decline), but employment in retailing also fell (by 16 pr cent) in response to the reduced economic base and lowered income that followed from Liverpool's new status and role in the national and international economy.

This changed status and role has been accompanied by a marked shift in the gender division of labour. While the restructuring of the docks in particular meant less work for men (with sea transport and dock work accounting for 47 per cent of the 29 per cent decline in male full-time employment in services), women's employment in services grew. But this growth was predominantly in part-time jobs. Women's employment in services only grew by some 3 per cent (around 3000 jobs) – about one tenth of the national rate. But, as Figure 6.3 shows, this growth was the net result of a decline in full-time work of 11 per cent (around (12,000 jobs) and an increase in part-time work of just over a quarter (around 15,000 jobs). The service sector remains the dominant employer of women in Liverpool, accounting for 83 per cent of total female employment in 1984. Three key sectors stand out: health and education (where 51 per cent of the jobs are now part-time), hotels and catering (especially pub and bar work where 75 per cent of the jobs are part-time), and

SERVICES

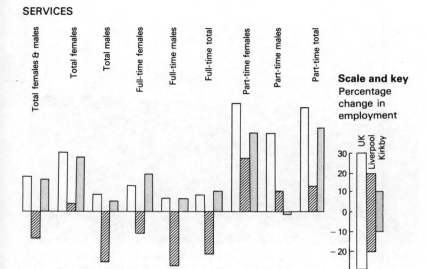

Figure 6.3 Percentage changes in employment in services: Britain, Liverpool and Kirkby by males and females, full-time and part-time, 1971 and 1984. Source: Annual Census of Employment. Calculated on the basis of the percentage changes between 1971 and 1984 (1968 SIC) and between 1981 and 1984 (1980 SIC).

retailing (especially mail order where 55 per cent of the work is part-time).

In the outlying areas there was some job growth in services. In Kirkby, for example, service sector employment rose by 15 per cent, but from a very low base (involving only some 1200 jobs). As in Liverpool as a whole, the overall changes again showed notable gender differences with women's employment increasing by over a quarter – nearly seven times the rate of growth in male employment. Three-quarters of the growth was in women's jobs which were again mainly part-time (60 per cent) and concentrated in 'miscellaneous services' (particularly pubs and clubs), retailing and education.

Behind this changing pattern of employment were a range of distinct restructuring processes in both private and public services. The spatial form taken by this restructuring bore some marked similarities to that in manufacturing. As in the latter, there was considerable closure of 'capacity' particularly in retailing. Some outer estates, including both Halewood and Speke, had never seen the development of major shopping centres and large shops. But those that had were badly affected by this retrenchment. Woolco, Asda,

Plate 6.2 An engineering factory in Knowsley Industrial Park with manu-facturing activities concentrated on part of the site and the remainder leased to a storage company.

Boots and Littlewoods pulled out in rapid succession as local income fell in Kirkby. Local banks have closed for the same reason. Again, as in manufacturing, there has been a concentration of service capacity. Vernons, the football pools company, closed its Halewood operations and now busses the retained workers to its Aintree and inner city sites. Its main competitor, Littlewoods, closed its Old Swan facility and transferred the work to Crosby. New technology has also been important in the restructuring: for example, Littlewoods has introduced new telephone and computerized order-ing and dispatch systems in its mail order business, and Barclaycard, in Kirkby, has brought in new computer systems to push up productivity. The growing use of part-time workers to give 'numer-ical flexibility' is particularly marked in the new private sector services that have filled some of the gaps left in the outer industrial areas by the closure of manufacturing plant. In Speke, a retail park of predominantly cash-and-carry centres has been set up. Operations here are conducted largely by a skeleton full-time staff with a larger contingent of part-time workers to meet peak business, chiefly at weekends and on one or two evening openings. In Kirkby, some of the disused industrial sites have been converted into distribution depots, but these offer relatively few new jobs.

The changed economic and political climate of the last decade has also presaged a major restructuring of public services. Again, there are marked parallels with the restructuring of private manufacturing and service sectors, particularly in terms of the 'rationalization' of service provision and the 'intensification' of work. Schools and hospitals have been closed in response to both the falling population levels (resulting especially, as will be argued below, from the outward migration that the deteriorating local economy has encouraged) and the increased restrictions on public expenditure. As in manufacturing, there are signs that those who have retained their jobs are having to work much harder and in increasingly stressful conditions. The signs are particularly clear in the local health service where attempts to meet the needs to provide competitive in-house tendering have resulted in reductions in the hours worked and the pay of ancillary staff (see Meegan and Pearson, 1989, forthcoming).

The outer estates of Merseyside in the spatial division of labour of the 1980s: labour market breakdown and growing impoverishment

The local labour market is disintegrating under the stresses engendered by the restructuring of the last ten to fifteen years. It is increasingly incapable of providing jobs in adequate numbers for those seeking them and is riven by a growing polarization in the pay and conditions of the work that is available – a divide marked, on the one hand, by the increased dominance of a reduced number of large employers and, on the other, by the desultory growth of small firms offering relatively low pay and little security.

The recent restructuring has left an ever growing lump of long-term unemployed in its wake. Registered unemployment for the estates as a whole averages nearly three times the national rate and nearly four times in some areas (such as Tower Hill in Kirkby):

'You never see anyone going to work in the morning.' (30-year-old Kirkby woman describing Tower Hill. Interview, 1986)

And what is particularly alarming is the pace at which young people are being added to the unemployment register.

'He's never had a job and wouldn't know how to go about getting one.' (45-year-old Kirkby woman talking about her 20 year old son. Interview, 1986).

Figures produced by Knowsley Borough Council (which takes in both Halewood and Kirkby), for example, show that despite a 27 per cent fall in the numbers of young people of school-leaving age since 1979, the total number of young people on the unemployment

Plate 6.3 Restructuring: from large to small companies. A former large factory broken up into small units.

register or involved in the temporary employment schemes that have burgeoned in the area in recent years has stayed remarkably constant. Only 7 per cent of school-leavers in 1987 in Knowsley found jobs (compared with 28 per cent in 1979). The registered unemployment rate for young people has effectively been kept down by a combination of 'schemes' and youngsters 'staying on' at school.

As the opportunities for full-time employment contract, the local labour market is polarizing between the large private and public employers and a heterogeneous collection of small firms. An indication, for example, of the changed fortunes of the industrial estate in Kirkby since its heyday in the second half of the 1960s is provided by the fact that while it now contains more than double the number of companies in operation then, this increased number employs only a third of the peak workforce of the 1960s.

The large companies that have kept restructured production in the area have found their dominance of the local labour market much increased – in terms of both the disappearance of former competitors for labour and the changed industrial relations climate engendered by the massive job losses of the recent restructuring. Recruitment has been drastically reduced, but on the rare occasions when these firms do recruit, they can, of course, be highly selective. Many do so from lists of workers already made redundant by them or already working for them on a part-time basis and even, in one case that we came across, workers who had recently gone into retirement. There is also some pressure from the workforce for 'kith and kin' recruitment, the attractiveness of which is heightened for firms by the depressed state of the labour market, since general advertisement of vacancies tends increasingly to produce what many companies describe as virtually unmanageable numbers of applications.

Pay and conditions in the small firms that have been established in the outer areas cannot match those of the large multinationals and there are signs that 'wage competition' for them is with state-determined levels of benefit and, increasingly, pay and 'allowances' associated with special employment and training measures. They are also able to exploit the 'psychic income' that young people attach to full-time jobs in a market dominated by special employment measures. Far from setting a base line for pay, the levels of remuneration on schemes are consistently 'underbid' by firms in this part of the labour market. The local careers offices have numerous examples of young people leaving training schemes for jobs paying less but seen as 'real jobs':

'They always end, they don't carry on . . . they're just schemes, not a job.' (18-year-old Kirkby girl just completing two years of schemes. Interview, 1986).

In this way, then, the downgrading of Merseyside's standing in the international and national division of labour has been reflected in a downgrading of its labour market, which has meant in turn a drop in income for the people living on its outer estates. A survey in Kirkby in the winter of 1981, for example, found that one half of the households in the sample were receiving supplementary benefit, 20 per cent were having rent and/or fuel bills paid by the state, 70 per cent claimed that they had no money left after paying basic bills, and 50 per cent felt that they were going without necessities in terms of food and clothing (CES Limited, 1982).

While it is difficult to put a precise quantitative measure on it, there is no question that the situation has got much worse since then. A recent analysis places Knowsley at the bottom of a 'living standards league table' of local authorities (CES Limited, 1988). Rent arrears in the borough have continued to rise, after a particularly sharp increase between 1979 and 1981 (when they rose by 40 per cent). In Halewood two thirds of tenants receive housing benefit while in parts of Kirkby, the corresponding figure reaches 81 per cent. For the borough as a whole, the proportion of schoolchildren being provided with free school meals has increased from 55 per cent in 1979 to just over 77 per cent eight years later. And the reports of local counselling and advisory bodies all confirm this bleak picture.

Behind these statistics are stories of people living in increasingly difficult and desperate circumstances:

'I'm always skint because I'm unemployed. I get depressed because I can't provide for my family. We have fights . . . It would be nice to have a few bob to spare. The only relaxation I can afford is a bottle of rum occasionally. I have ideas about DIY but have to scrimp to buy wood. We get paid every two weeks; my wife and I work out to the penny what we're going to spend it on. A lot of people are in this situation. It's quite common. (36-year-old unemployed Halewood man, studying part-time at Halewood Comprehensive school. Interviews, 1986.)

'We haven't had a holiday since he [the husband] became unemployed [7 years] . . . . The kids get a holiday with the school or the community centre. It gives them a break from the area. When your giro comes every fortnight, you have to set aside so much for food, gas and the leccy, and there's very little left. But there's always something extra – kids' clothing or money for house redecoration – and you never get the extra, so you have to delay paying the gas and leccy bills. Obviously we can't afford to run a car and rely on catalogues for the children's clothes and ours. Christmas time is the worst. You just can't buy things at Christmas.' (Kirkby married couple, the wife 32 and the husband 34, both out of work with three dependent children and living with a retired parent)

Socio-economic change in Outer Merseyside: the role of 'locality'

The economic restructuring of the outer estates: the 'local effect'

The economic restructuring described in the previous section was not a simple, 'top-down' process of management-instigated change and workforce compliance. While international and national macro-economic pressures dictated the need for production reorganization, this reorganization had to be pursued in a particular local context which influenced the precise form that the restructuring finally took, the way in which it operated, and the responses that it both encouraged and had to accommodate.

Changing spatial divisions of labour and the recomposition of the workforce.

Merseyside's service role in the spatial division of labour of early industrialization, based on the docks and related activities, bequeathed a local workforce steeped in 'casualism' (Charles and

Meegan, 1989). To the planners, the new waves of manufacturing investment and the decanting of Liverpool's population to the outer estates offered the possibility of creating a new role for Merseyside in a spatial division of labour based essentially around mass production. But there could be no clean break from the old to the new. The new labour market had first to coexist with, and then evolve from the old. It was never an easy coexistence. The new mass production factories drew labour selectively from the small companies in the inner cities, driving up wages in the process. The building of the outer estates did the same in the construction industry with large companies using, for the first time on any large scale, factory-based construction techniques. The competition between the 'old' and 'new' employers was not always won by the latter. The headmaster of the first purpose-built comprehensive school in Kirkby, for example, commented rather ruefully on the way that the school's building programme had been interrupted: the carpenters working on it left the site when a ship docked in Birkenhead for refitting. Such was the custom – the work in the shipyards was irregular but relatively well-paid and interesting, and was therefore gladly seized whenever the opportunity arose.

On balance, however, the terms of labour market competition were heavily biased in favour of the new mass production factories. Irregularity of work was replaced, for a time at least, by regularity, but the price of this transition for local workers was more discipline and supervision at work and relative loss of control over the labour process. The old traditions were being tested by the management attitudes and methods of the new employers, especially the multi-nationals. While this new 'testing ground' offered wider opportunities for labour organization, as the trade unions took root in the new factories, it was still a harsh one with which to come to terms. A workforce with an inbred 'come day, go day' attitude towards organized work, suspicious of management motives, used to some degree of control over their work processes and less fearful of, or at least inured to periods of unemployment, had to contend with a new breed of employers demanding 'malleability' and compliance to the dictates of 'Fordist' mass production.

A qualitatively unique segment of the labour force has emerged from this interaction which could be said, at the risk of caricature, to display a cynicism towards the 'system': a cynicism which is, on the one hand, sustained by a sense of 'fair play' and a history of mutual support, while being tempered, on the other, by the aptitude for moulding the 'system' to its own purposes. Signs of this remaking of the Merseyside working class can be seen in the corporate restructuring of the last ten or fifteen years and in the labour force's response to it.

The local experience of economic restructuring: resignation, resistance and 'working the system'

Many local workers feel bitter about their experience of this restructuring. Here is how one worker at Dunlop's Speke factory expressed his feelings, in a letter to the company chairman on the announced closure of the factory:

'I have worked at Speke for eighteen years as a maintenance fitter and . . . I feel I must make it known to your Board just what it means to be told by the firm that you have served for half a lifetime that you are no longer wanted.

When I first went to Speke the factory worked six and sometimes seven days a week and I worked twelve hours a day and even longer at weekends . . . I can honestly say that during those years I cannot remember my four children growing up.

In recent years I have watched the place slowly crumble away, both the plant and the moral fibre of the workforce. I don't pretend to understand economics or industrial strategy. I only know about the reality of my own work situation. May I take the liberty of quoting from a report I wrote for an Action in Work group [a Christian Fellowship group] I belong to. . . . 'During my years at the factory discipline has become less and less, no one seems to want to work any more, the policy seems to be do as little as possible for as much as you can get, and I am afraid that I have become as guilty as the next man. We are constantly being told to pull our socks up, that the company will not go on forever losing money in our factory, but it seems to be all talk, nothing is ever done about it. It seems to me that the company are deliberately allowing the plant to get into this situation so they have an excuse to close it down' . . .' (Ray Turner, 1979).

Experiences such as these only served to harden workforce scepticism, fostered in the days of casualism, towards employers. This cynicism, however, did not prevent attempts at resisting the restructuring. Indeed, in a number of cases this resistance appears to have been undertaken on principle, as a protest against its perceived injustice and breach of 'fair play' – irrespective, it often seemed, of personal cost. Thus, for example, the workers who went on strike at the Speke Dunlop factory did so even though this meant losing the enhanced redundancy payments offered by the company for 'going quietly'. This protest was built upon the local tradition of mutual support as, for example, in the 'community pickets' of the Standard Triumph and Dunlop factories in Speke, or the 'sit-ins' in Kirkby at Commonwealth Curtains, IPD, Massey-Ferguson and Plessey.

This sense of 'fair play' is a powerful one. One manager of a private training establishment in Kirkby, for example, commented on the way in which his training staff would 'walk out on principle' if they felt that they were being unfairly treated – regardless of local job prospects. In our discussions with local employers a theme that kept cropping up was the publicly perceived intractability of the local workforce. Most were adamant that this did not mean that the workforce was unmanageable, only that a much more direct management approach was necessary: 'they're a tough lot but if you can get them on your side they're world beaters'; 'we have our differences but they're in the open and we sort them out'. Several of the companies in the 'modern core' could boast that they had very few major disputes and that, although different in the ways suggested by the above quotes, industrial relations were no worse than at any of their other plants elsewhere in the country.

'Working the system' or 'beating the system' were phrases repeatedly mentioned in our interviews with employers, workers and trade unionists – generally to refer to the workers' predilection for devising their own strategies for complying with management instructions, most notably in the struggle over the control of work processes. While not unique to the workers on Merseyside, the general view seemed to be that these tactics were more highly developed in the area because of its history of insecure employment. One example of 'working the system' that we came across was in response to reorganization of work and 'speed up' in an engineering factory. The pace of the work was set by a time study engineer on the basis, as one worker described it, of 'the hardest job on the line . . . so the woman with this job has the hardest time'. The response by workers in some sections of the factory has been to set up their own informal system – 'the girls set it up that they swap and everyone does a turn on the hardest job for a set period of time'.

Voluntary redundancy in an area of mass unemployment and low income

The response to restructuring, however, was not always one of organized resistance either through the formal confrontation of strikes and sit-ins or the informal tactics of 'working the system'. In many instances, and to the surprise of both management and trade unionists, there was little resistance to plant closure and job loss. But why, in an area so seriously affected by unemployment? The answer again seems to lie in the area's economic and social history.

With a history of insecure and relatively poorly paid work, many found the offer of a one-off lump sum redundancy payment impossible to refuse – and, just as with casual work, there is no time for equivocation, the offer has to be seized and 'tomorrow will take care of itself'. A former worker at the Standard Triumph plant in Speke remembered how the management offered enhanced redundancy payments before a mass meeting of workers was held to discuss escalating industrial action. The new offer was particularly attractive not just because it involved payments above the statutory figure, but also because it brought in workers who were not eligible for statutory payments because they had worked in the factory less than 2 years – which, given the relatively high labour turnover of the plant, was a significant number. And, for management, it had the desired effect:

'Before the meeting the men were saying, '£1000 – not to be sniffed at, that' – when they were on £3000 a year. Edwardes [Michael Edwardes, the company's managing director] had planned the turnaround in the men's attitudes. Like a street trader, he'd started off with the lowest bid and then increased it, to show [to the trade union leaders] how the workforce attitudes would change.

I was amazed at how they turned on 'Red Robbo' [Derek Robinson, the leader of the shop stewards' combine committee] when he recommended turning down the new offer. They were easily led. They just fell for the dangling carrot. The whole place was in uproar, they were calling him all sorts of names – 99 per cent of them. And the previous meeting they'd all backed him to the hilt. It was just greed. They saw the money and wanted it – but it was nothing really.'

We raised the issue of workers opting for redundancy in a discussion with a group of shop stewards at Ford's Halewood plant, which had seen a major voluntary redundancy programme in 1983. The stewards stressed that this programme had to be judged against the backcloth of 'life in the plant'. When the management had first asked for volunteers there had been no takers, but the offer was soon 'oversubscribed'. The reorganization of production played an important role.

With the major changes in working methods that accompanied the introduction of new technology, especially in the body shop, the company had shifted into a period of what seemed like 'continual change', a shift which, together with public questioning of the plant's future, had raised 'a big, dark cloud over the factory'. Workers were being moved around the factory into areas with different working traditions, with unfamiliar shop stewards, for example, or where they were no longer eligible for overtime, or where the move meant a loss of 'seniority' (the trade unions operated an informal system of 'job progression' tied to specific areas of work). Workers affected in this way often felt themselves to be 'outsiders' in their new work areas and their pride suffered accordingly. For others, especially those working on the dayshift without overtime, low pay was also a factor – many of these workers were already receiving Family Income

Supplement, for example. Given the increased pace and pressure of work, why not 'take the money and run'?

The take-up of redundancy was influenced, then, by monetary considerations (especially, of course, where what was at issue was the complete closure of a plant – the lump sum payments were attractive enough in themselves, so why turn them down if the plant was going to go anyway?) and by work stress, intensifed by the restructuring of production. Another factor that appears to have been important had its roots outside the world of work in 'civil society' more generally – patriarchy.

As we argued earlier, the new waves of manufacturing investment of the 1950s and 1960s brought women workers into regular full-time and part-time paid employment in significant numbers for the first time. When the restructuring of the 1970s started to produce job loss, many fought it. Indeed, in some plants women workers took a leading role (as at Plessey's factories on the outer estates). The fact remains, however, that substantial numbers of these new workers did opt for redundancy when it was offered. Work changes were important, of course, as were the redundancy payments themselves, but the attractiveness of the latter was increased by a local culture that was deeply patriarchal.

Local teachers and youth workers testify to the persistence of the old attitude that a 'woman's place is in the home' (it is 'nan' who looks after the kids); if her place is in a job, then it is in a 'woman's job'. One teacher that we spoke to saw gender stereotyping as a major problem in her careers work at a Kirkby comprehensive, and to counter this has organized conventions at the school focusing on women in 'non-traditional' jobs: she accepts that it is still 'an uphill struggle'. The leader of a local youth workshop agreed with her, deploring the fact that, with a few notable exceptions, they have had relatively little success in interesting young women in work traditionally done by men and that, in the workshop, 'it is the girls who head for the kettle' still.

Gender divisions remain entrenched at work. One woman we spoke to who worked on an engineering production line remarked ruefully that it was the women who did the work while the men 'sit around gabbing' or 'kicking paper balls around'. There were no men on her line although the line setter was a man. Up until five years ago the shop steward for her section had been a man even though women outnumbered men 10:1 and he had never worked on the line. There were some signs of change (not least because of the growing number of women who are now sole wage earners in their households) and the number of women shop stewards has increased, but, she argued, these women still find meetings outside work difficult, having to cook evening meals and fulfil other family commitments. It is not surprising, therefore, that she felt that what few disputes there had been at the factory had all been 'soft little things for men' (individuals being disciplined by management, for example) and not over women or women's issues.

The patriarchal nature of the local culture thus helped ensure that the foothold that women held in the labour market, after the investments of the 1950s and 1960s, was always very tenuous. In these circumstances redundancy payments were often seen by women as an unexpected 'bonus' for curtailing a life in paid employment that had been beseiged from the start by social pressures tying them to the home.

Restructuring, workers' cooperatives and an 'enterprise culture'

Another way in which the response of the 'locality' to the restructuring of the 1970s has been conditioned by the cultural legacy of previous rounds of investment and the labour markets that these helped to create can be seen in the much-publicized demise of worker cooperatives, and in particular of KME in Kirkby. This remnant of the 1960s wave of investment in domestic appliances (it started out as a branch plant of Fisher & Ludlow, itself a subsidiary of the British Motor Corporation), was for five years between 1974 and 1979 the largest producer cooperative in Britain. Its eventual collapse was a bitter disappointment for the cooperators:

> 'We had the chance to change society. It's something this country needs. We did offer it. We've been let down. We've let ourselves down in many ways.' (Jack Spriggs, Worker Director, KME Ltd. Quoted in Eccles, 1981).

The reasons for this failure were complex, including hostility from politicians and the civil service, economic conditions, and organizational factors. But what seems to have been particularly important was the conflicts of interests and ideology that participation in the venture imposed on the workforce (see, for example, Eccles, 1981; Tynan and Thomas, 1984). Here was a workforce conditioned begrudgingly to the 'gift' of employment in the 'Fordist' factories of the 1950s and 1960s, with their highly developed division of labour and heavily supervised routine work, being presented with the task of cooperative self-management. The transition could never have been trouble free:

> A shop-floor culture which is collective, oral, reactive and which deduces from experience is quite different to a managing culture which is indi-

vidualistic and takes initiatives based on analysis . . . (Eccles, 1981, p. 377)

'We're used to fighting the management. It is hard for us and our members to realize that we are the management'. (Shop stewards quoted in Eccles, 1981, p. 391.

It is difficult to shake off this cultural legacy. It also surely helps explain why there has been no flowering of formal 'enterprise culture' out of the soil laid and cultivated by 'Fordist' manufacturing industry. A recent study by the MSC's North-west Regional Manpower Intelligence Unit provides some revealing figures. At the time of the 1981 Census of Population, Knowsley, of all the local authority districts in the north-west, had the lowest proportion of self-employed people in its labour force. Its figure of just over 6 per cent was, for example, well below that of Bolton (12 per cent), Allerdale in Cumbria (12.3 per cent) or Blackpool (20 per cent). More recent figures on the take-up of the Enterprise Allowance Scheme confirm this relative lack of 'entrepreneurship'. In the 1985/6 financial year Knowsley had by far the lowest ratio of 'starts' on the scheme, with only 1 for every 97 eligible takers. The next lowest was Copeland in Cumbria (1:54) and the highest St Helens (1:28). And local economic development officers confirm the difficulties they have found in attempting to generate an 'enterprise culture' based on small firms in the area.

Locality response: emigration, 'the leaving of Liverpool' and its outer estates

Given the severity of the economic restructuring that it has experienced, it is no surprise to find Merseyside now leading the metropolitan counties in the rate of its population decline. In the first half of this decade the region was losing population at an annual rate of 7 per thousand, nearly twice the rate of the next largest loser (Greater Manchester) and five times the figure for London (Britton, 1986). Against this backcloth of regional decline, the former recipients of population dispersal – the outer estates – are themselves now declining. Over the 1970s, population loss ranged from 7 per cent in Speke to 31 per cent in parts of Kirkby, with the lower rate in Speke reflecting the fact that decline had set in earlier in the 1960s (it is the oldest estate) while others were still growing.

There does appear to have been a major shift in the pattern of migration from Merseyside and its outer estates over the last twenty years or so – a shift closely linked to the recent rounds of economic restructuring. In the early 1970s the moves were predominantly 'housing led' to the New Towns of Runcorn and Skelmersdale just over the county's boundaries (Wood, 1985). Since then, although it is difficult to measure precisely, the balance has shifted towards moves further afield in search of work. Some families have clearly left the area for good but there has also undoubtedly been a growth in the importance of people 'working away' from Merseyside for brief spells of time.' But at what social cost? As one 17-year-old Kirkby woman put it; 'You can't have a weekend Dad.'

Community under siege: the making of an 'underclass'?

Emigration for most, however, is neither economically feasible nor socially welcome. For these people a living has to be sought, as best as can be achieved, in the local labour market. There is no question that the recent bout of economic restructuring and the associated labour market fragmentation have set up great pressures in the 'social fabric' of the outer estates. There are certainly signs of widening divisions between the employed and the unemployed, and within the former between those employed in the relatively established large firms and those in low paid insecure work.

It is the workers in the large firms who have tended to buy their council houses (or who live in the private housing developments dotted around Kirkby and Halewood); it is they who own more cars and use these increasingly to shop and socialize outside the outer estate town centres (contributing of course, to the downward spiral of local shopping centres in a process of 'cumulative causation'). On the other side of the divide are the low-paid and unemployed, increasingly dependent on the state, trapped in the lower-quality housing stock, reliant on public transport and tied to the declining local shopping facilities (declining that is, in terms of the quality and range of goods on offer, if not their price). Marginalized in the labour market, this group is also increasingly marginalized in the social life of the estates.

Although it is difficult to quantify, there is undeniably a pronounced sense of alienation amongst the 'marginalized' groups in the estates. It expresses itself in a variety of ways. Antipathy towards the state, both national and local is one:

'You belong to the state, you are run by the state. That's how it feels in Kirkby.' (A Kirkby man interviewed in the *Daily Mail*, 27 March 1985).

'You're going to find there's two big problems round here – the "Council" and the "Social"' (A male single parent with three young children, temporarily lodging, with his sister – also a single parent with six children – while trying to be housed. Inteview, 1986)

Drug abuse could be interpreted as another way in which the alienation of some, especially the young, shows itself. While the extent of drug abuse on the outer estates is in no sense of 'epidemic' proportions, there is still no question that it has increased in importance in recent years. One recognition of this is Knowsley Council's current sponsorship of a drugs education unit based in Kirkby, complementing previous initiatives of Liverpool City Council and the Merseyside County Council. Vandalism has also been a longstanding problem on the estates, with public buildings and especially schools providing the main targets for this expression of juvenile discontent.

Indeed, what struck us particularly markedly in our research was the pronounced degree of mistrust that many groups on the outer estates feel towards 'outsiders', a mistrust reinforced by the alienation from, and antipathy towards figures of 'authority' be they DHSS 'snoopers', local authority housing managers, officials from the gas or electricity boards, or even 'researchers'. One major 'qualitative' effect of the recent economic and political restructuring that these areas have undergone is undoubtedly the creation of this climate of fear and suspicion. While not wishing to exaggerate the extent to which this 'siege mentality' has taken root in the outer estates – when we were able to establish our 'credentials', local people responded with an exceptional generosity and openness to our questioning – the fact remains, that among many of the marginalized groups and throughout a substantial part of the 'urban and regional system' there is a growing resentment towards being 'investigated'. What kind of society is it that forces, for example, groups of its citizens to circulate photographs, in the form of 'Wanted' posters, of its civil servants (in this case, DHSS claims investigators)?

Given the undeniable economic and social stress that people on Merseyside's outer estates are having to face, the question arises whether they can be said to constitute what has come to be described as an 'underclass' (see, for example, Auletta, 1982 and Dahrendorf, 1987). Such a class is generally distinguished by its experience of a combination of various forms of disadvantage – unemployment, poverty, low levels of education and literacy, 'unstable domestic circumstances' – and its inability to cope with this situation. As the previous discussion has shown, evidence of all these different forms of disadvantage can be found on Merseyside's outer estates but, and this is an important qualification, there is also substantial evidence of an ability on the part of the residents of these areas to cope with this adversity in a socially coherent and cohesive manner. If an 'underclass' is being formed in these areas is is not without a substanital degree of resistance.

Community under siege: resilience and resistance

Voting patterns in local council elections would certainly caution against an uncritical acceptance of the creation of an undifferentiated 'underclass' on the outer estates. In Kirkby, for example, turnout at local elections has in fact increased as the local economy has deteriorated, from an average of around 27 per cent in the mid-1970s to some 38 per cent in 1987. The increase occurred, moreover, in all of the area's wards (and not just those containing, for example, the relatively more prosperous private housing estates) – hardly the response of an 'underclass' disenfranchising itself from mainstream politics.

There are other signs of a community resisting the dislocation of social life. Despite their appallingly poor employment prospects, children still turn up regularly for school on the estates and still generally behave responsibly while there. Thus, for example, while attendance figures at Kirkby's Ruffwood Comprehensive have fallen in the last three years to 82 per cent, they are still equal to or above some of the equivalent figures of relatively more prosperous times in the mid-1970s. And even in the fifth forms with the worst prospects of employment or further education, three out of four still regularly attend school. Children on the estates are resisting being pushed into an unemployed subculture and, in their dress and manner, show how they want to participate in the broader society on equal terms with their peer groups elsewhere in the country.

Nor are the outer estates the centre of crime that some media images might suggest. Kirkby, for example, has been saddled with the 'lawless' image since the TV series 'Z Cars' in the 1960s but the statistics do not support this. Vandalism remains a problem but there are signs that it is on the wane, and the latest figures show that the number of burglaries has declined in recent years. Cases of murder and sexual crime are particularly low. Merseyside's outer estates, despite their serious economic deprivation, have not become, by any stretch of the imagination, dangerous places in which to live.

The resilience of the locality also shows itself in a number of 'bottom up' initiatives in which local people have joined together to try to improve their living standards and quality of life. These initiatives include credit unions, unemployed centres, housing cooperatives and, in the case of Speke, an active women's 'health action group' that has campaigned, with notable success, for improved public health care facilities on the estate. An important distinguishing feature of the initiatives is the active role played in them by groups of people whose economic and social circumstances might, in localities with a different class profile and different cultural traditions, leave them marginalized in community life, prime candidates for membership of the 'underclass'.

To understand this resistance to social dislocation it is necessary to appreciate local social, cultural and political factors that an all-embracing term like 'underclass' misses and which together combine to produce a local 'uniqueness' that in turn helps to condition 'locality response'. A number of aspects of this 'uniqueness' in relation to Merseyside's outer estates can be singled out. First, there is the remaking over time of the family support networks that were broken by the initial population dispersal from the inner city. There are now families of three generations living on the outer estates. In Kirkby, for example, at the beginning of the 1980s about one half of the 'households' had never lived anywhere else and 60 per cent (and 80 per cent in some areas) had close relatives living near by. Around two thirds of adults in Kirkby had either parents or children in the area, and about a half had brothers and sisters living there (CES Limited, 1982). The young children of 'Bunnytown' are now parents themselves and their parents are now grandparents.

Another feature of the area that emerged particularly strongly in our 'household' interviews was the very close identification our interviewees had with their 'locality'. Concern was variously expressed by the people we spoke to over the possibility that we might 'talk down Kirkby' (or Speke, or Halewood). Most of the people we spoke to said that, despite the fact that the area had 'gone down' noticeably in recent years, they still wanted to stay:

> 'Kirkby is a good place and getting better as it gets more established. The kids of the people who first came here from Liverpool have married and are grown up. They're Kirkbyites. Going to Liverpool is an adventure.' (43-year-old Kirkby man, unemployed. Interview. 1986)

One unemployed 17-year-old used her attachment to family and Kirkby to justify her decision not to take the 'dole express' down to Wimbledon where her uncle had promised to find her full-time work:

> 'I can't leave, I love my home. I have my freedom and I'm very happy. It would be no good if I went down there anyway – I'd be so unhappy I'd just depress everyone.' (Interview, 1986)

An older woman, also unemployed, summed up her feelings towards the area, and the reasons why she would never consider leaving, in a typically 'scouse' way:

> 'If a fellah's van breaks down, everyone passing will try to help. They'll probably make a balls-up of it like, but they help you!' (Interview, 1986)

The sense of humour of the Merseyside working class has been all but mythologized in discussion of the area but there is no doubting its existence – especially when one is on the receiving end of it as we so often were in our interviews! A mixture of self-deprecation and the teasing put-down of authority (or 'outsiders' and 'woollybacks'), it is clearly an integral part of the local culture of 'vitality amidst fatalism', an oft-quoted phrase which could also be used to describe some of the things that we were saying earlier, in the context of industrial relations, about 'working the system', acting 'on principle' and labour force scepticism of management motives. The humour draws on its own vocabulary of truncated words – a world of 'scallies' and 'divis' 'robbing the leccy' – and the expressive turn of phrase. One couple described the area in Kirkby where they lived – an area with the highest unemployment levels – in the telling phrase: 'the saying is "they bury their own dead in Tower Hill"'! Its origin lies in the custom of door-to-door collections for funerals – which, given the relative youthfulness of the area's population (it was Kirkby's own 'overspill' estate for the children of the first settlers) can be sad occasions. But what it conveys particularly well is the sense of 'community' that has been forged on parts of the estates.

And this sense of community does mean that the economic divide referred to earlier between the employed and the unemployed does not simply translate into a political or social one. There remains a deep awareness of, and resentment towards, the opening up of these divisions. The employed have direct experience of unemployment if not through their family then through their neighbours. For many the financial burden of helping relatives in difficulty is compounded by the added pressures and worries of being the 'lucky' ones with jobs amidst a growing number of unemployed, pressures and worries engendered by a range of emotions from deep-felt empathy with the plight of the latter to the perceived need to protect property from theft or vandalism. One working woman put her finger on the feelings of most of the employed people we spoke to when she queried the purpose of our survey by asking directly, 'Is it to help the unemployed?' Another, working part-time as a telephonist in Kirkby with a husband also in part-time employment, one son is in full-time employment and another unemployed, all living on a local private housing estate, was particularly concerned about the growing isolation of the elderly and young mothers caused by the rundown of the local shops:

'You don't get to know them because you don't see them. So you don't know if they're OK.' (Interview, 1986).

This identification with, and concern about, the unemployed by the employed certainly qualifies the argument that recent changes in class structure are marked by a growing trend towards the social distancing of these two groups, with the unemployed forming a detached category of 'untouchables' (Newby et al. 1986). If such a process is at work then it is surely subject to a great deal of regional differentiation – it may hold true, for example, in Cheltenham, but it is the argument of this chapter that it most certainly does not in Kirkby. Schools on the outer estates, for example, are increasingly attempting to cater for the educational needs of older unemployed people. Halewood Comprehensive runs a scheme in which unemployed or retired people in the locality can attend 'O' and 'A' level classes alongside their children or grandchildren. The success rate has been very encouraging and currently there are over 80 adults with ages ranging from 18 to 80 studying at this 'community school' – an initiative which shows one way in which the local community is refusing to accept the 'marginalization' of the unemployed.

Unemployment is no stranger to Merseyside and its outer estates and arguably no longer carries with it the social stigma that it retains in other parts of the country. Consequently, there is a much wider participation of the unemployed themselves in the social life of these areas than in localities where unemployment is less pervasive. One indication of this is provided by the recent growth of local authority-supported housing cooperatives in Kirkby. One scheme is run by a group of young women, all single parents, currently living in particularly rundown housing conditions. All the schemes (there are seven in operation with more in the pipeline) involve unemployed people. Indeed, in one cooperative three quarters of the participants are unemployed. In another, the members have included in their plans additional housing for elderly and disabled people not in the founding group but to be recruited from the council's housing list – a gesture of inclusion that speaks volumes for the local sense of community.

One of the most obvious signs of the confidence of local unemployed people in their own abilities to organize a collective response to their situation is the establishment and growth of unemployed centres. The first union branch in the country mainly for the unemployed came out of the economic restructuring of 'outer Merseyside' – after the closure of the Standard Triumph car plant in Speke in 1978. A network of unemployed centres throughout Merseyside has subsequently developed. Of this network, it is generally agreed that the most active and firmly based centre is the one in Kirkby. The centre provides a counselling service for local unemployed people and is particularly effective as a 'go-between' for

Plate 6.4   Cherryfield Housing Cooperative starting on site in Kirkby in 1986.

those intimidated by the bureaucracy of the local state. It actively campaigns over such issues as claimants' rights, fuel costs and disconnections, and the encouragement of women to 'sign on': it estimates, for example, to have been responsible for the take-up in a three year period of nearly £1 million of benefits that would have remained unclaimed but for its intervention.

Why is all this happening in Kirkby? We have already referred to strong attachments to the area by local residents, the family-based support networks and the refusal of some of the unemployed to be marginalized in the social life of the locality. In the case of the unemployed centres, the local history of political activism also appears to have played an important role. This political activity focused in the early days of the estate on struggles over housing conditions and rents, and the experience of these struggles laid the foundations for organization around the unemployed centre. The unemployed centre has a policy of maintaining direct links with local workplaces through the trade unions and has provided direct (and politically controversial) support for local workers in industrial disputes. This link between 'civil society' and 'production' is an important one politically and is exemplified in the current campaign on Merseyside known as 'One Fund for All'. Some 30,000 workers throughout Merseyside currently contribute to this fund which is used to support unemployed centres throughout Merseyside and, by linking funding directly to trades unions, aims to overcome, or at least minimize, the political uncertainties and controversies of local government support. The workers at the unemployed centre in Kirkby are currently partly financed by the Fund. In this way, then, political struggle and organization have also been important in conditioning the form and nature of 'locality response' to economic restructuring.

Politics and policies: 'local effectivity' in the response to restructuring

Local government in the area has also been involved in political struggle in recent years, being increasingly handicapped in its attempts at ameliorating the worst effects of restructuring – being caught in the scissors of declining local resources (as restructuring cuts into the industrial and commercial rate base, rent arrears rise and emigration and falling population reduce entitlement to rate support) and the increasingly stringent restrictions imposed on local government spending in the last eight years by the Conservative central government. Although both of the local authorities responsible for the estates we studied (Liverpool for Speke and Knowsley for Halewood and Kirkby) are Labour controlled, they have responded to this situation quite differently (Meegan, 1989, forthcoming).

As social and economic conditions have deteriorated, there has been a radicalization of the Labour party on Merseyside in general and especially in Liverpool. After nearly a decade in which political control in the city swung between Labour and Liberal/Conservative coalitions, a radical Labour group was elected in 1983. This was reflected in policies, with a shift towards a major municipal housing (and unemployment) programme. Speke became one of the 'Priority Areas' in the council's 'Urban Regeneration Strategy'.

This strategy involved the focusing of resources on housing and environmental improvements in some twenty-two areas of the city with particularly acute levels of poverty and deprivation. At the centre of this strategy was the building of semi-detached houses to replace the interwar tenements and the postwar high- and medium-rise blocks of flats and maisonettes. The funding of this ambitious programme in the context of central government restrictions on the authority's budget contributed to a dramatic but eventually abortive confrontation with central government. Forty-seven of the city's Labour councillors were surcharged and barred from public office.

The authority's confrontational stance and some of its policy emphases also created internal political divisions within the local Labour party, as well as alienating broad sections of the community (most notably over race relations and policies towards housing cooperatives). The newly elected Labour authority now faces a financial crisis as it attempts to meet its budgetary commitments (it is one of the seventeen local authorities 'ratecapped' by central government in the forthcoming financial year).

Housing has been a major plank in Knowsley's policies, too, but the political and policy emphases have been very different. The council has been more pragmatic in its relations with central government and has cooperated with it in its policy making. This has involved, for example, the sale of a housing estate formerly owned by the municipality to an independent trust; the introduction, with central government and EEC support, of 'community refurbishment' and environmental upgrading schemes; and, more recently, the sponsoring of cooperative housing ventures. But even with this approach the council has not been able to avoid some internal political divisions over spending cuts in the funding of voluntary sector initiatives (most notably the unemployed centres), produced in part by the need both to keep within budget constraints and to fall in line with the shifting emphasis in the central government's urban programme funding towards 'economic' as opposed to 'social' projects.

The regeneration of Merseyside's outer estates

This chapter has attempted to show how Merseyside's outer estates have ridden the roller-coaster of Britain's postwar economic growth – from the halcyon days of the 1950s and 1960s to the current retrenchment and decline. The estates, marginalized in the prevailing spatial division of labour in the UK, are now having to contend with a labour market unable to meet the employment aspirations of the local people. Some are having to move in search of work but, as the chapters on Cheltenham and Swindon show, the chances of breaking into the housing and labour markets of the more prosperous areas in the south are declining. For those that remain the future, in terms of the opportunities for regular, decently paid employment, will remain bleak in the absence of any major policy initiative. And the social fabric of the locality is going to be tested further.

Yet the problems that these areas face are not 'local' in origin. The economic restructuring that they have experienced took place on a global scale and the funding constraints on local initiatives are, to an increasingly substantial degree, imposed by central government. The solution to the problems must, therefore, also be sought at these levels – through a national-level policy towards urban renewal which builds upon local initiative, and an international programme of economic regeneration. In terms of the former, it is encouraging that a group of local authorities, all with economically and socially disadvantaged outer estates, have joined together to form 'Radical Initiatives for Peripheral Estates' ('RIPE') to lobby central government and to place the issues of outer estates firmly on the national political agenda. And it is a radical national programme of economic regeneration that is required. This chapter, it may be hoped, has demonstrated that local resources in Merseyside's outer estates – especially the collective strengths of the local people – are available on which such a strategy could build. Perhaps we could leave the last words to an unemployed Kirkby woman whose friend was attempting to move from Kirkby to Northampton:

'It's the reverse of Kirkby. There's lots of jobs and no houses. It's daft – here there are lots of houses and no jobs. Why don't they bring the jobs to Kirkby?'

References

Auletta K., (1982) *The Underclass*, New York: Random Books.

Britton M., (1986) 'Recent population changes in perspective', *Population Trends*, 44, Summer, OPCS.

Centre for Environmental Studies (CES) Ltd (1982) 'Kirkby: an outer estate', *CES Paper* 14, London.

Centre for Environmental Studies (CES) Ltd (1988) 'People and places: a classification of urban areas and residential neighbourhoods', *CES Paper* 33, London.

Charles, B. and Meegan, R. A. (1989, forthcoming) 'Politics, culture and mass unemployment: the case of Merseyside's outer estates', in M. Boddy et al. (eds), *Politics, Culture and Place* (provisional title).

Dahrendorf, R. (1987) 'The erosion of citizenship: its consequences for us all', *New Statesman*, 12 June.

Eccles, T. (1981) *Under New Management*, London: Pan.

Fletcher, B., Johnston, C., Ord, J., Prescott H., and Ritchie, H. (1982) 'Learning a living; the social context: an educational perspective for Halewood', mimeo.

Gentleman, H. (1970) 'Kirkby Industrial Estate: theory versus practice', in R. Lawton, and C. M. Cunningham (eds), *Merseyside: Social and Economic Studies*, London: Longman.

Lane, T. (1987), *Liverpool, Gateway of Empire*, London: Lawrence and Wishart.

Lister, P. H. (1983) 'Regional policies and industrial development on

Merseyside 1930–1960', in B. L. Anderson and P. J. M. Stoney (eds), *Commerce, Industry and Transport: Studies in economic change on Merseyside*, Liverpool: Liverpool University Press.

Liverpool City Council (undated) *Industrial Liverpool*, Report prepared for the Liverpool City Council by the Finance and General Purposes Committee under the Chairmanship of Alderman Sir Alfred Shennan.

Lloyd, P. E., (1970) 'The impact of development area policies on Merseyside 1949–1967', in R. Lawton and C. M. Cunningham (eds), *Merseyside: Social and Economic Studies*, London: Longman.

McMonnies, D. (undated) 'Trade unions and co-ops? A Merseyside case study: the Scott Bader Synthetic Resins saga', University of Liverpool, Department of Political Theory and Institutions Working Paper 6.

Meegan, R. A. (1988) 'Economic restructuring, labour market breakdown and locality response' in J. Morris, A. Thompson and A. Davies (eds), *Labour Market Responses to Industrial Restructuring and Technological Change*, Brighton: Wheatsheaf Books.

Meegan, R. A. (1989, forthcoming) 'Political and policy responses to economic restructuring in a beleaguered locality', in M. Harloe et al. (eds), *Place, Policies and Politics: Do Localites Matter?*, London: Unwin Hyman.

Meegan, R. A. and Pearson, M. (1989, forthcoming) 'Restructuring and the state: the health sector in a declining region', in J. Lovering and R. A. Meegan (eds), *Restructuring Britain* (provisional title).

Merseyside Council for Voluntary Services (1978) 'One in eight: a report on eleven outlying housing estates where one in eight of Merseyside's people live', Liverpool.

Newby, H., Vogler, C., Rose, D. and Marshall, G. (1985) 'From class structure to class action: British working class politics in the 1980s', in B. Roberts, R. Finnegan and D. Gallie (eds), *New Approaches to Economic Life. Economic Restructuring, Unemployment and the Social Division of Labour*, Manchester: Manchester University Press.

Roberts, J. (undated) 'Kirkby: a short history', Public Relations Department, Knowsley Metropolitan Borough Council.

Thompson, F. Langstreth (1945) *Merseyside Plan 1944*, London: HMSO.

Tynan, E., and Thomas, A. (1984) 'KME – working in a large cooperative: worker perceptions and experiences in the Kirkby Manufacturing and Engineering Company, 1974–79', Open University Cooperatives Research Unit, Monograph 6.

Wood, P., (1985) *Agenda for Merseyside*, Merseyside County Council, Development and Planning Department.

7

'Not Getting on, Just Getting by': Changing Prospects in South Birmingham

DENNIS SMITH[1]

Introduction

One locality, several 'communities'

'South-west Birmingham' is not a description that locals would readily use or recognize. It is the part of the city most clearly dominated by white residents predominantly employed in skilled and unskilled manual occupations. The label applies to a cluster of eight wards: Northfield, Selly Oak, Kings Norton, Brandwood, Weoley, Longbridge and (as a result of boundary reorganization in 1983) the newly created wards of Bournville (inserted between Selly Oak and King's Norton) and Bartley Green (carved out to the west of Weoley). About 200,000 people live there.

South-west Birmingham is the heartland of Birmingham's white working class. Asian and Afro-Caribbean families are far less common there than in the 'inner city'. Professional men and women generally prefer to settle in the suburbs of north Birmingham towards Sutton Coldfield and south-east Birmingham towards Solihull. Nevertheless, the locality is internally diverse. Within its boundaries are to be found private housing as well as council estates, 'yuppified' enclaves as well as clusters of deprived households. There are a number of distinct centres of social activity – including shopping centres, schools, surgeries and public houses – which provide recognized points of reference for a large number of 'communities' and overlapping social networks.

For example, the Pershore Road, one of the two main routes through the locality, passes through Selly Park, Stirchley and Cotteridge in the north, skirting the Bournville chocolate factory and the residential properties of the Bournville Village Trust. The estate was established by George Cadbury in 1900, avowedly with the purpose

[1] *Management School, Aston University, Birmingham*

235

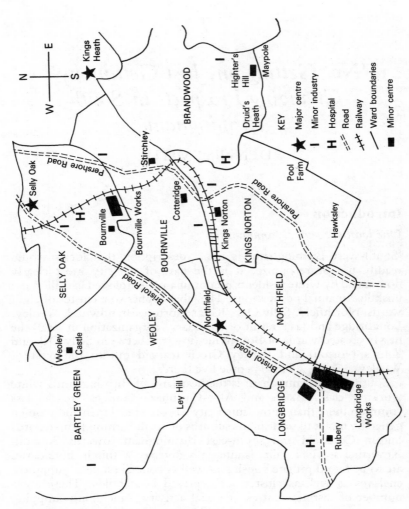

Figure 7.1 Sketch map of south-west Birmingham

of encouraging working men and women to lead decent and responsible family lives. It is still difficult to find a pub in the area. Ironically, Bournville is little more than a stone's throw away from the blue cinema and massage parlour on the Bristol Road.

The Pershore Road feeds into the Redditch Road which, in turn, passes the large post-war council house estate of Hawksley before crossing the city boundary going due south. Running roughly parallel to Pershore Road in the northern part of the locality, the Bristol Road leads into the very busy shopping centre of Northfield. Two miles further down the same road, at the south-western extremity of the locality, is the Austin Rover car factory at Longbridge.

There is a high degree of continuity in the development of south-west Birmingham. From the beginning it offered scope for people who wanted to do well for themselves. The ancient parishes of Kings Norton and Northfield beyond Edgbaston Park and the city boundary constituted a kind of southern frontier for those seeking to enhance their fortunes in the late nineteenth and early twentieth centuries. By 1840, the three canals cutting through these fields had been supplemented by the Birmingham and Gloucester Railway. Stations were opened on its route at Kings Norton (1849) and Northfield (1869). In 1876 the Birmingham and South-West Suburban Railway was opened, taking passengers down to Selly Oak, Bournville and beyond from the middle of town. Within three years the Cadbury brothers had opened their new factory. Just over a quarter of a century later, a further railway station was opened three stops down the line from Bournville at Longbridge. In that same year, 1905, Herbert Austin opened his car works.

In the period before World War I the chocolate and car factories both played an important part in attracting new residents into south-west Birmingham. Cadbury acquired a national reputation for cleanliness and care. Both the factory and the village (legally a quite separate concern) were bathed in a romantic glow in the reports of visiting journalists:

The employees, mostly girls of all ages, from thirteen upwards, have smiling, rosy faces, and they put into their work the ardour that they would into a game of marbles. (*Daily Mail*, 20 September 1906)

[Mr Cadbury] has built the model village of Bournville, provided swimming ponds for the male and female employees, recreation grounds, cricket and football fields, lawns and shrubberies, gymnasia and reading rooms, besides educational facilities of the most varied description. Indeed, the happiness and health of the workers have ever been the first consideration of the firm. (*Yorkshire Herald*, 26 September 1906)

Agents for Herbert Austin were drumming up labour in many areas outside Birmingham. By 1918 he had built an estate of 252 dwellings 'on garden suburb lines' complete with village hall, mission rooms and steam laundry (Church, 1979, p. 43). However, Austin did not attempt the ambitious programme of community betterment favoured by Cadbury. The car manufacturer concentrated upon attracting unskilled workers who would find his wages acceptable. His style was autocratic and individualistic. He admired Henry Ford and was much more overtly hostile to trade unions than were the management at Cadbury. Other firms joined the drift to south-west Birmingham. For example, land was purchased in Northfield by Morland and Impey (a firm which later became Kalamazoo Business Systems) in 1900 for £5500. Thirty-seven years later it was revalued at £47,000, one indication of the increasing popularity of the area (Smith, 1964, pp. 196–70).

Residential growth complemented industrial expansion. Skilled engineering workers, anxious to escape the inner-city slums, came out to smart terraced houses with small gardens in Selly Oak, Bournville, Stirchley, Cotteridge and the streets close to the stations in Northfield and Kings Norton. In 1911 the city boundaries were expanded to include most of Kings Norton and Northfield Urban District. By 1918 substantial parts of the wards of Selly Oak and Bournville, as currently defined, had been built up. Most of the eastern part of Northfield ward had been developed as had been the Bristol Road to the west of North-field village. Between the two World Wars there was rapid development closely related to the expansion of the car industry. The inhabitants of large new housing estates south-west of Kings Norton village 'were drawn there largely by the opportunities for work provided by the Austin motor-car works at Longbridge' (Elrington and Tillott, 1964, p. 21). Large estates also sprang up in Weoley Castle – 'ambitiously planned . . . (with) numerous *culs-de-sac*, curving roads, and the provision of central community buildings' (Tomlinson, 1964, p. 56) – and still further to the south-west beyond Northfield village in the direction of Longbridge.

By the late 1930s a very large proportion of car workers were getting to work on wheels. A survey of methods of travel by employees at Longbridge works and another factory at Castle Bromwich (to the north-east of Birmingham) showed that in 1937 only ten per cent were walking to work. Many more were travelling by train (8.5 per cent), bus or coach (25 per cent), private car (14 per cent), motor cycle (2 per cent) and bicycle (13.5 per cent) (West Midland Group, 1948, p. 69). This suggests that the close association between factory and residential community cultivated at Bournville was less in evidence at Longbridge.

Since the Second World War several more huge council house estates have been built to the west and the south. Kings Heath and Kings Norton have been extended to the city boundary. Longbridge is reaching towards the Lickey Hills, a popular beauty spot just south of the city. A major extension to the west has occurred, pushing across into the far reaches of Bartley Green. Much of this development was presided over by the Labour administration on Birmingham city council which held sway from 1952 to 1966.

Recent tendencies

During the two decades since the mid-1960s four tendencies have manifested themselves in the development of south-west Birmingham. First, a significant proportion of the Austin workforce have, like their Cadbury counterparts, developed strong roots in the locality. Recent interviews with car workers residing in the area, both young and old, revealed that a high proportion had lived locally for large parts of their working lives. Many of the younger men had lived there all their lives. By the 1960s a number of families who had built up a modest prosperity were aspiring to improve themselves by buying their council houses or moving onto a private development. At the same time they were experiencing pressure from below. This pressure resulted from a second major trend, originating from before the war, which was the council policy of rehousing occupants of substandard inner city accommodation on new suburban estates. For example, Druids Heath Estate in the south of Kings Norton ward was built in the mid-1960s partly to cater for families made homeless by the clearance of slums in Balsall Heath. By the early 1980s the composition of the various wards, by occupation and housing, was as shown in Table 7.1.

Selly Oak, in the northern part of the locality, stands out as a ward with high house ownership and a relatively high proportion of residents with professional, managerial and non-manual occupations. As an example of a ward at the other extreme, Longbridge, in the south-western part of the locality, has a much lower representation of 'top jobs' holders, a significantly higher proportion of skilled and semi-skilled manual workers, and a greater reliance on council housing. In fact, with the exception of Selly Oak (12 per cent) and its near neighbour Bournville (16 per cent), the share of housing owned by the city council varies between wards within the approximate range of 30–65 per cent. In these wards, between thirty and forty per cent of council dwellings are flats. High levels of council building have recently occurred in Kings Norton and Bartley Green.

A third tendency evident in the development of south-west Bir-

Table 7.1 *Socio-economic groups and housing types in south-west Birmingham*

	SO	Bv	Nd	Bd	Wy	BG	KN	Lb
Socio-Economic group (%, 1981)								
Professionals, employers, managers	14	14	12	10	9	8	11	6
Non-manual	30	31	30	30	27	26	26	24
Skilled and self-employed	22	21	24	22	25	26	25	26
Semi-skilled	19	21	23	22	26	25	24	28
Unskilled	5	6	5	5	7	8	7	6
Housing type (%, 1983)								
Council	12	16	31	42	51	63	60	47
Owners	68	57	60	50	35	32	37	44
Flats	20	32	36	40	34	39	38	35
Houses	72	58	54	52	59	47	52	54

SO = Selly Oak, Bv = Bournville, Nd = Northfield, Bd = Brandwood, Wy = Weoley, BG = Bartley Green, KN = Kings Norton, Lb = Longbridge
Source: Ward profiles, Development Department, Birmingham City Council.

mingham has been that the proportion of Asian and Afro–Caribbean residents has very gradually increased, most noticeably in the northern part of the district. For example, between 1971 and 1981, the number of residents born in the New Commonwealth and Pakistan increased from just under 2500 to just over 4700. South-west Birmingham remains overwhelmingly (well over 95 per cent) white although Asian labour from residential areas such as Selly Oak has tended to concentrate in specific industrial niches, for example the Cadbury night shift. Although direct statistical evidence is not available, it has been estimated that between a third and a half of night workers at Bournville are black.

Finally, the pre-war movement of industry-led residential growth in south-west Birmingham has, in some respects, recently been reversed. In the past, the city council has been under pressure to build new homes for people flocking into that part of the city to take up readily available jobs. More recently, however, the need has been felt to encourage new businesses and investment capital to move into south-west Birmingham to provide jobs for people who are already there. One recent political initiative of this kind has been the opening in November 1986 of the Woodgate Business Park in Bartley Green, adjacent to the newly designated Woodgate Valley Country Park and close to Junction 3 of the M5 motorway. The city council's Development Department has spent over £2 million on the business park which is 'designed with new technology based firms in mind'. The council's Economic Development Unit is intending to offer users of the business park 'advice and support on property development, financial and employee training requirements' (*Birmingham Post*, 13 November 1986).

The switch from a situation in which large amounts of labour are being pulled in by industrial capital to one in which the city council is seeking to pull in industrial capital to soak up excess labour (and improve the rate base) is a profound one. For example, the following newspaper report would not have appeared in the 1960s:

Rumours about the possibility of vacancies at Austin Rover's Longbridge plant led to a local Job-centre being besieged yesterday by hundreds of unemployed. Dozens of men queued in the snow and rain. . . . By early afternoon, more than 400 men and women had visited the centre, all asking about jobs at Austin Rover. (*Guardian*, 4 January 1984)

Within Northfield one of the largest new buildings is occupied by the DHSS offices. Twenty years ago, few local people would have thought that such establishments would become as significant as they now are in the lives of the communities making up south-west Birmingham. In the period 1948–69, when the average UK unemployment level was 1.3 per cent, Birmingham returned a level of only 0.9 per cent. South-west Birmingham was the heartland of blue-collar affluence in a region which accepted prosperity as a fact of life. In 1965 the West Midlands produced more gross domestic product per head of population than any other British region. Incomes per household in the West Midlands were 13 per cent above the national average, according to the Family Expenditure Survey 1961–3 (*Sunday Times*, 27 February 1983; Sutcliffe and Smith, 1974, pp. 169–70).

The 1980s provide a harsh contrast with the 1960s. The West Midlands has suffered so severely in the recent recession that it is now ranked alongside regions such as Northern Ireland and Scotland which have a much longer experience of economic misfortune. By 1981 unemployment had risen to above 10 per cent in all the wards of south-west Birmingham. In Longbridge ward about one in seven (15.3 per cent) of all economically active residents were out of work. Five years later, the relevant proportion was about one in five, not only in Longbridge (19.5 per cent) but also in adjacent Northfield ward (21.4 per cent). During the late 1970s and early 1980s the number of hourly-paid employees at Austin Rover's Longbridge plant fell from nearly 19,000 (in 1977) to just over 11,000 (in 1982). During the next five years they fell by a further thousand. To take another key case, in 1980 the Bournville factory owned by Cadbury-Schweppes employed about 7000 shopfloor workers. By the middle of the decade, that workforce had been reduced by approximately half.

Two major local employers

Austin Rover and Cadbury-Schweppes (or 'Austin' and 'Cadbury' as they are more likely to be referred to locally) are the largest manufacturing employers located in south-west Birmingham. Excluding office staff, they provide about 13,500 jobs between them. According to a post code analysis of the Longbridge hourly-paid workforce, in 1986 over 40 per cent lived in the seven postal districts (B14, B29, B30, B31, B32, B38, B45) which cover the area. A similar analysis of Cadbury shopfloor workers showed that about 65 per cent of male employees and 80 per cent of female workers lived in south-west Birmingham. About 1500 workers, or over a third of the local Austin contingent, are clustered in the B31 postal district which includes the whole of Northfield ward and a sizeable chunk of adjacent Longbridge. Nearly a thousand, or about a quarter of the local employees, reside in B45 district which fans out to the south-west and north-west of the plant itself.

In contrast to the overwhelmingly (97 per cent) male shopfloor workforce at the car factory, the chocolate factory shopfloor splits roughly equally between men and women, the former having a slight (55:45) numerical superiority. About half the total shopfloor workforce live in the postal districts B30 (in which the factory is situated), B31 (directly to the south-west) and B29 (immediately to the north and north-west). The direct influence of Cadbury and Austin Rover as local employers is most strongly felt within a broad swathe of residential estates extending outward from the Bristol road along its length from Selly Oak in the north-east to Rubery in the extreme south-west. However, this influence also reaches out towards Weoley and Bartley Green in the north-west and towards Kings Norton and Brandwood to the south-east. It is noticeable that while the Longbridge plant recruits fairly evenly throughout south-west Birmingham outside of its immediate environs, Cadbury has particular drawing power in Kings Norton and Brandwood, wards which have a number of council house estates along their southern perimeters. The firm has recruited about one in six of its current female workers from postal districts B38 and B14, which include large tracts of the two wards just mentioned. A random sample of twenty-five Cadbury women residing in these districts showed that eighteen were part-timers. Of these sixteen were married and all but four were born in 1940 or earlier.

Plate 7.1   Typical housing in the Longbridge area.

About 4200 residents of south-west Birmingham work on the Longbridge shopfloor. Another 2000 or so are on the Cadbury payroll in a similar capacity. In 1981 just over 84,000 residents of the relevant wards in the area had jobs. Obviously, a proportion of those in employment work outside the immediate area. Although the outer boundaries of the postal districts of the wards of south-west Birmingham do not coincide, it is possible to make a very rough estimate that shopfloor jobs at Cadbury and Austin Rover account for about one thirteenth of *all* occupations held by residents of south-west Birmingham. Account should also be taken of the many white-collar jobs on the two sites and the dependence of subcontracting firms (such as Boxfoldia in Bournbrook and, in Kings Norton, Triplex Safety Glass Co. and Burman and Sons) on regular orders from Cadbury and Austin Rover.

The reduction in jobs at the two firms in the late 1970s and early 1980s was very rapid and, in its scale, completely unprecedented. In both cases, the redundancy programme was preceded by direct confrontation with the existing local trade union leadership. This phase included the highly publicized sacking by the Longbridge management of Derek Robinson, principal union convener and an articulate opponent of the particular recovery strategy favoured by Michael Edwardes, the chief executive of British Leyland. In each

243

Plate 7.2   South-west Birmingham housing of the 1960s.

case, major attempts have been made to link massive investments in new machinery with widespread revision of working practices which increase the control of management and elicit greater flexibility and cooperation from the workforce. Although these processes began in the latter years of the last Labour government, the strategy of industrial restructuring accompanied by payroll reductions was from the start strongly encouraged by and has become closely associated with the present Conservative government. A high proportion of the jobs lost in the two firms were obviously held by local residents. If the workers made redundant from Austin Rover and Cadbury in the late 1970s and early 1980s were representative of the total shopfloor workforce in their residential patterns, then about 5000 south-west Birmingham people from the two firms underwent that experience during those years. Those remaining with the firms have had to endure a stressful process of change which is far from over.

The processes of restructuring Austin Rover and Cadbury-Schweppes, in particular the transformations occurring at Long-bridge and Bournville, are to some degree representative of changes occurring within the city and the wider West Midlands region. In the second section of this paper, this broader context of change will be considered. Following this, the specific processes occurring at the two factories will be examined in a little more detail.

In the third section, evidence of voting patterns in south-west Birmingham and the wider area will be considered. Subsequently, the patterns of industrial restructuring the electoral behaviour identified will be located within a broader context. Some distinctive characteristics of south-west Birmingham as a locality during the 1980s will be discussed with particular reference to their implications for the shaping of politics and policy.

Dimensions of restructuring

The impact of restructuring, redundancy and recession since the early 1970s has varied, depending upon whether you are referring to the West Midlands as a whole, the Birmingham area, south-west Birmingham or the particular cases of Longbridge and Bournville. First to be considered will be the case of the region as a whole.

The scale of recent economic decline in the West Midlands was recently summarized by Geoff Edge, chairman of the West Midlands Enterprise Board:

> Unemployment in the West Midlands climbed from under 6 per cent in 1979 to over 15 per cent in 1985, the highest increase for any region in the UK. Half the unemployed have now been out of work for over a year. Since 1978 over one third of all manufacturing jobs have been lost. During 1984 alone over 1300 companies failed in the region and 10,000 redundancies were announced in the Metropolitan County areas.

He went on to complain that during the 1970s the West Midlands had 'the lowest investment per head of any of the older industrial regions in Western Europe.' (WMEB, 1986).

The regional strategy review produced by the West Midlands County Council Economic Development Unit in 1985 had painted a similar picture, pointing out, for example, that in the sphere of vehicle production and metal manufacturing, one worker in eleven had been made redundant in 1982. Of the manufacturing jobs lost in the West Midlands between January 1980 and June 1983 42 per cent were in motor vehicles and mechanical engineering. Since 1982 most of these losses had been in the components sector. Restructuring at Longbridge certainly made an impact since it has been estimated that in the early 1980s about 60,000 jobs in the metropolitan county depended on Austin Rover (WMCC-EDU, 1985, pp. 6, 61).

The loss of manufacturing employment has to some extent been offset by an increase in service sector employment. About 200,000 manufacturing jobs were lost to the West Midlands between 1966 and 1979; over the same period about 150,000 jobs appeared in the service sector. However, since 1979 the expansion of service sector employment in the West Midlands has occurred at a much slower rate than the loss of manufacturing jobs.

The general pattern just outlined – a shift from manufacturing towards service sector jobs within the context of an overall decline in employment and investment – encompasses important variations at the other three levels mentioned. This can be explored, in part, by comparing tendencies in the 1984 Birmingham travel-to-work area as a whole and, within that area, in south-west Birmingham in particular. The latter is covered by employment offices in Selly Oak (which deals also with Edgbaston and Harborne), Northfield (extending south to Barnt Green) and Kings Heath.

First, it is useful to compare employment trends in metal and engineering industries and service industries. South-west Birmingham (SWB) had continued to be (slightly) more dependent upon the former and (significantly) less upon the latter than Birmingham as a whole. Between 1971 and 1981 the proportion of metal and engineering jobs within the total fell from 42 per cent to 30 per cent in SWB (declining to 27 per cent by 1984); from 37 per cent to 29 per cent in Birmingham (24 per cent by 1984). Over the same period service sector employment increased from 40 per cent to 56 per cent in SWB (rising to 59 per cent by 1984); from 53 per cent to 62 per cent in Birmingham (65 per cent by 1984) (DoE sources). However, over this period SWB has been able to compensate for losses in metal and engineering by increases in service sector jobs to a greater extent than Birmingham as a whole. Specifically, Birmingham lost over 177,000 jobs in metal and engineering (43 per cent) but gained only about 26,000 service sector jobs (an increase of 7 per cent); SWB lost nearly 16,000 metal and engineering jobs (42 per cent) and gained over 11,000 service-sector jobs (an increase of 30 per cent).

Within these overall trends, there are important shifts with respect to the balance between male and female jobs and full-time and part-time employment. The changing composition of the workforce can be expressed by tracing the number per thousand employees who fall into each category. Over the decade from 1971 to 1981 the proportion of SWB workers who were male and full-time fell from 625 per thousand to 540 per thousand (declining to 531 per thousand by 1984). In Birmingham as a whole the fall was less steep – from 596 to 555 (falling to 531 by 1984). During the same period the proportion of female part-timers rose in SWB from 131 to 204 per thousand (203 per thousand in 1984). In Birmingham as a whole the rise was, again, less steep – from 135 to 170 (180 in 1984). These statistics are compatible with the suggestion that employees in the south-western part of the city are experiencing in heightened fashion (relative to the Birmingham TTWA as a whole) the pressures associated with increased demands for a 'flexible' labour force, much of it female and part-time.

It is interesting to notice that in Birmingham as a whole employment rates in the two remaining categories – male part-time and female full-time – were approximately the same in 1971 (27 and 242 respectively) and 1981 (29 and 245). By contrast, in south-west Birmingham while male part-time employment remained almost static between the two dates (252 in 1971, 241 in 1981), female full-time employment increased from 131 to 219 per thousand. Over the decade the degree of participation in the labour market by females from SWB wards increased from just under 40 per cent to just over 40 per cent: for example, from 39 to 42 per cent in Brandwood and Kings Norton. This suggests that women are increasingly being drawn into accepting 'second best' occupational choices: for example, women who would ideally like to work part-time may find themselves having to take up full-time work for economic reasons.

Within the manufacturing sphere, south-west Birmingham suffered a greater percentage loss of jobs in motor vehicles than Birmingham as a whole, losing 23 per cent of male full-time jobs as opposed to a 15 per cent loss in the larger area. To some extent these figures reflect the fact that expansion of production facilities was occurring during this decade outside south-west Birmingham in areas such as Solihull where British Leyland was working on a new international luxury car project which did not finally collapse until the early 1980s (Whipp and Clark, 1986). In the period 1981–4 employment in this sector fell still further, both in SWB and Birmingham as a whole. During these years the larger area lost about 18,000 jobs in the car industry. About 2500 of these were in south-west Birmingham.

Within the SWB service sector, there was a reduction of employment in the sphere of retail and distribution between 1971 and 1981. With the exception of the sale and distribution of food and drink, where its decline was roughly equivalent to Birmingham as a whole, SWB retail and distribution employment declined to a much greater degree than Birmingham: by 11 per cent as opposed to 5 per cent. The reason for this difference is unclear but, in fact, it was offset to a slight degree by the fact that although the number of recorded jobs in public houses declined by about 5 per cent in Birmingham as a whole, in SWB pubs the figure increased by over 20 per cent (the number of female part-timers more than doubling). It is fascinating to speculate about the reasons for these trends. Were those people (the vast majority) still in work prudently spending their money on entertainment locally rather than going abroad or even 'into town'? Did it become easier to recruit barmen and barmaids? In the period from 1981 to 1984 retail and distribution employment began to pick up slightly in SWB while remaining almost unchanged in the larger area.

In contrast to trends in retail and distribution, jobs in education and also the medical and dental sphere increased disproportionately in south-west Birmingham between 1971 and 1981. In particular, female employment in education, both full-time and part-time, increased by over 200 per cent during the decade (although some of these gains were lost in the early 1980s). A fair proportion of these jobs were, no doubt, taken by teaching staff moving into south-west Birmingham from other parts of the West Midlands or beyond. Nevertheless, many of the less skilled jobs were probably filled by 'locals'. The presence of a number of NHS establishments in south-west Birmingham helps to account for the buoyant market for female part-timers in the medical sector. The number of such jobs increased by over 130 per cent over the decade, reaching a plateau in the early 1980s.

In conclusion, compared to Birmingham as a whole, the south-western segment lost a larger proportion of its core manufacturing employment during the 1970s and was less able to generate a large increase in the proportion of service sector jobs. There is a further point. The pattern of job loss at the Cadbury-Schweppes factory at Bournville was such that in that decade male full-timers and female part-timers both fell by over 40 per cent. This cut across the tendency for the latter to increase while the former fell. On the other hand, the large investment programmes that were agreed for Cadbury and Austin Rover at the end of the 1970s represented an important exception to the generalization that it was difficult to bring about such programmes in the West Midlands. At this point it will be useful to examine in more detail the context in which capital was being put into, and labour squeezed out of, the two companies just mentioned.

Mergers and markets

During the last quarter of a century employees at both 'the Austin' and 'Cadbury' have seen their respective parent companies become, initially, larger and more complex and, subsequently, smaller and more streamlined. The streamlining has produced a paradoxical effect. On the one hand, employees have had to endure massive local cut-backs in labour requirements. On the other hand, the relative significance of Bournville and Longbridge within the two companies has apparently significantly increased. In 1984, Cadbury-Schweppes were employing approximately 17,000 people in the United Kingdom with another 16,500 scattered in North America, Australia, New Zealand, Africa and Asia. About 9000 of the UK employees were shopfloor operatives. In the same year Austin Rover's 38,000 workforce included about 25,000 shopfloor employees, mainly at Longbridge and Cowley. Each of the companies drew over a third of its national shopfloor workforce from south-west Birmingham.

In fact, in each case the local factory has had a long association with the company at the heart of its affairs. Longbridge was the original headquarters of the Austin Motor Company established by Herbert Austin in 1905. The factory at Bournville was built by Richard and George Cadbury on a green-field site in 1879. Both establishments became parts of much larger operations through a series of mergers.

Austin merged with Morris Motors (based at Cowley) in 1952 to form the British Motor Corporation (BMC). In 1965 BMC purchased Pressed Steel, a major independent body supplier and joined with Jaguar to form British Motor Holdings (BMH). The British motor industry was by that time divided between BMH and Leyland, a bus and lorry manufacturer who had absorbed Rover. Three years later, BMH merged with Leyland to form the British Leyland Motor Corporation (BLMC). More recently, British Leyland (later BL, later still Austin Rover, now the Rover Group of British Aerospace) has contracted through the loss of Jaguar, which became independent in 1984, and the sale of the Leyland trucks division which merged with Daf in 1987, the Rover Group retaining 40 per cent of the new company. By 1987, Longbridge and Cowley, the basis of the Austin-Morris axis of 1952, were once more the key sites within the company.

Following a merger with J. S. Fry of Bristol, Cadbury established overseas operations between the wars in Canada, Australia, New Zealand, Ireland and South Africa. In the postwar period Cadbury made steady profits but by the early 1960s the company was looking for ways to strengthen its market position as American competition increased. In 1964 Cadbury bought Pascall and Murray, two confectionery firms, and began to diversify into cakes, tea, instant mashed potatoes and other food and grocery products. Five years later Cadbury merged with Schweppes, a bigger company specializing in carbonated drinks which had recently taken over Typhoo Tea of Birmingham. Cadbury-Schweppes was reorganized into five divisions: confectionery, drinks, food, tea and coffee, and overseas operations. The strategy of product diversification pursued during the 1970s was being considerably modified by the early 1980s. However, in 1985 the range of products was still fairly broad. Within the UK, apart from group functions (employing 722 people on four sites including the London headquarters), there were four divisions as follows: confectionery, employing 6739 on 4 sites; drinks (3576 on 8 sites); beverages and foods (4930 on 5 sites); and health and hygiene (1249 on 5 sites).

By 1987 two divisions – beverages and foods, health and hygiene – had been sold off. Meanwhile, the soft drinks side of the business had been strengthened by the acquisition of Canada Dry and the establishment of a new joint venture with Coca Cola. Cadbury-Schweppes had declared its intention to 'concentrate on those businesses it knows best – confectionery and soft drinks' (*Guardian*, 5 September 1986). At the end of 1986 there were nearly 7000 employees (of all kinds) in the confectionery division compared to about 3000 in the drinks division. The confectionery division was spread between three other sites, apart from Bournville. These were at Somerdale (near Bristol), Chirk (near Wrexham in North Wales) and Marlbrook (located roughly midway between Somerdale and Chirk). The Bournville site currently remains the most important of the confectionery sites. It contains two major operations (in the 'moulded' and 'assortments' factories) as well as a large office complex servicing the confectionery division as a whole.

Restructuring, redundancy and technological innovation

The circumstances leading to restructuring at the Longbridge and Bournville plants will now be considered in more detail. At the time of the merger between Austin and Morris in 1952, European producers were still recovering from the war, American manufacturers were preoccupied with the home market, and the British government had been encouraging the export efforts of British firms. The British car industry had been for over five years 'one of the country's most capable foreign currency earners' (Dunnett, 1980, p. 32). However, during the 1950s government pressure to export relaxed. Firms such as BMC found that they could make good profits selling to the British market. Home sales grew by nearly 40 per cent a year in the early 1950s. As international competition increased, the British contribution to total output by the major car producing nations declined: from 11.4 per cent (1960) to 8.5 per cent (1970). Export share almost halved in the 1960s, falling to 14 per cent (Dunnett, 1980, p. 118). During the 1970s and early 1980s foreign imports made great headway in the UK market. The proportion of this market captured by vehicles produced in the UK declined from 72.3 per cent in 1973 to, 32.4 per cent in 1983 (Jones, 1985, p. 163). The proportion has risen since 1983, according to the company.

As foreign competition grew, mergers occurred (BMC, BMH, BLMC) and government involvement gradually increased. The culmination of this process was the 'rescue' of BLMC in 1974–5. The government purchased 95 per cent of the shares in BLMC, which was renamed British Leyland, and invested substantial amounts of public money in order to promote rationalization and technological change within the company.

Despite attempts to implement proposals made in the Ryder Report (Ryder Committee, 1975), by 1977 productivity had not significantly improved and BL's market share had fallen from over 30 per cent (1975) to under 25 per cent (1977). Since 1977 two chairmen and chief executives have had a major impact: Michael Edwardes who stayed from October 1977 to September 1982, and Graham Day, whose appointment was announced in March 1986. Four aspects of restructuring which have had noticeable effects at Longbridge will be mentioned.

First, there have been joint ventures with the Japanese firm Honda, for example in the development of the Rover 800 and the R8. The latter, which is to be built at Longbridge (*Birmingham Post*, 19 December 1986), is intended to come on stream at the end of the 1980s, filling a niche in the market currently occupied by the Maestro and Rover 200 models. This programme is intended to improve the marketability of Austin Rover cars but it raises questions of two kinds in some workers' minds. One relates to the future viability of power-train production (including engine and gear boxes) in Austin Rover, especially in view of Honda's own developing facilities at Swindon. Since August 1986 it has been announced that Austin Rover's 'K series' of advanced engines will be produced at Longbridge. The company states that 400 new jobs have resulted from such developments, including work on the PG1 gearbox and new small car gearbox. According to one power-train employee (speaking in August 1986), 'Since going in with Honda a lot of jobs have gone.' A colleague in the same area feared that there would 'probably be a big clear out'. In fact, research and development work on new engine and gear box facilities continues at present.

The other question concerns the relative significance of Cowley and Longbridge in the development and production of the new model range being planned by Graham Day's management team. A couple of comments from Longbridge employees make the general point:

'[Longbridge] is important. It's a big modern factory. They've spent millions on it. It must be important or it wouldn't be here.... Its importance has increased because of other plants closing. New cars have been brought here such as the Rover [200]' [Mini track worker]

'We only get rumours about how well they're doing at Cowley. We accepted change more readily than Cowley. There's a feeling that we only get the bottom end of the market [at Longbridge]. We would have liked the Maestro or the Montego – a bigger car that's wholly British-built. They've got the fleet market cars. They've got newer buildings down there. . . . We're led to believe that our food is better than Cowley's but I don't know how true that is.' (Indirect worker, New West works)

Whether or not the perceptions revealed in these two comments are accurate, management may well benefit from the feelings of rivalry that they express. In fact, it was announced in April 1987 that the R8 would be built at Longbridge.

A second aspect of restructuring to be considered is the part played by new technology. Installation of the Metro line gave Austin Rover a lead in automated car body production. Furthermore, parts of the Longbridge operation which are unprofitable to maintain, such as the foundry and forge, are being shut down. Foundry work has been transferred to Beans Foundry in Tipton: all Longbridge foundry workers have been offered jobs there, while forge workers were offered jobs elsewhere at Longbridge. However, productivity increases may have contributed to the creation of substantial over-capacity relative to market demand. A recent study which states the issue rather harshly argues that 'On the Metro line, the company invested more than £100 million in creating inflexible excess capacity which cannot now be used to do anything except build an extra 200,000 Metros that nobody wants to buy.' (Williams, Williams and Haslam, 1987, p. 62). It is possible that if sales of current models had been higher it might not have been necessary to enact a recent plan to cut 1265 white-collar jobs (supervisory as well as office staff), including about 380 at Longbridge and Drews Lane, Birmingham (*Birmingham Post*, 31 January 1987).

A third aspect of restructuring has been the attempt to achieve more flexible and continuous use of labour on the shopfloor. This has occurred in two phases. The first involved a fairly direct attack by management upon local union rights, especially the 'mutuality' arrangements which had allowed individual shop stewards to bargain over specific job times. Established tea-break and 'togging up' time allowances were also undermined. During strikes (for example, in November/December 1981) management wrote direct to workers, bypassing the union. A harsher regime developed at the level of line supervision. The second phase began in the mid-1980s. Its main features were: a shopfloor attitude survey conducted by outside consultants; the introduction of zone circles and line briefings in order to improve communication at shopfloor level between the foreman and his group; more elaborate induction programmes for new

Plate 7.3   Workers' cars outside Austin-Rover's Longbridge factory.

recruits including 'family days'; and a greater emphasis upon training. This new approach, partially reminiscent of Japanese practice, has the objective of eliciting the commitment of workers to the company and its products. It seeks voluntary cooperation in the pursuit of high quality and high productivity.

A final aspect of restructuring has been the redundancies. These mainly occurred in the early 1980s along lines set out in Michael Edwardes's 1979 'recovery plan'. Although the hourly-paid labour force was stabilizing at about 10,000 in the mid-1980s, rumours of Graham Day's own 'recovery plan' revived redundancy fears in early 1987. For example, the sale of Leyland Trucks was expected to be followed by over 2000 job losses, though not at Longbridge.

How does the case of Cadbury-Schweppes compare? In 1977 and 1979 Bournville experienced its first major strikes for more than two decades. The 1977 strike concerned wages and brought out the whole shopfloor. Its occurrence hardened the resolve of management to reconstruct the bureaucratic and time-consuming system of worker participation at Cadbury. One object was to weaken the influence of senior shop stewards by decentralizing consultation and bargaining. Management pressure in this direction was closely associated with attempts to introduce new working patterns. The 1979 strike was primarily concerned with management's desire to introduce a new four-shift pattern of working. The strike was costly for the company which decided to defer this innovation until later when very favourable financial incentives were offered to the workforce. Such a deal was necessary in order that investment in new plant to produce Wispa could be fully exploited.

The Wispa bar played an important part in the £120 million programme which got under way in 1980. By 1985 £10–£15 million had been spent on new plant for Wispa which required only 25 per cent of the labour force necessary on more conventional machinery. Launched, and then relaunched after initial problems, the reported Cadbury view was that 'Wispa is as important to us as the Metro . . . was to car makers Austin Rover' (*Sunday Times*, 6 January 1985). However, Wispa technology, like Metro technology, was capable of making far more of the product than the market could bear in the mid-1980s. In July 1985 Wispa workers were facing pay cuts of 'up to £30 a week' as the plant moved from a seven-day continuous shift pattern to the six-day pattern followed elsewhere in the factory (*Birmingham Evening Mail*, 23 July 1985).

Among the lessons which some managers feel they have learned from the Wispa project (and others) are: the need to minimize the social isolation imposed upon certain workers by capital-intensive systems; the value to management of increasing worker involvement in monitoring highly automated machinery; the gains to be made from getting production workers to carry out routine maintenance and cleaning tasks; and the art of balancing the advantage of 'team spirit' against the problems posed by 'elitism' within or among work groups. More flexible team-based production is being encouraged, including some experiments in 'annual hours' schemes which reduce overtime costs. Some of this thinking is being embodied in new plant currently being installed.

Returning to the comparison with Longbridge, management at both factories are attempting to minimize demarcation problems among craft and maintenance workers, striving for a simple distinction between electricians who are mechanically competent and mechanics who also have basic electrician's skills. In both cases, training programmes are being developed to enhance this strategy. Bournville have also made some experiments with 'product improvement groups' just as experiments with 'zone circles' have occurred at Longbridge. However, in both cases there had been only limited acceptance of these initiatives by management and unions by the middle of 1987.

Acquiescence of the workforce in about 3000 redundancies at Bournville was made easier by relatively generous financial arrangements that were offered within an 'agreed voluntary redundancy'

(MVR) scheme. Some key union representatives took the money early on. Much of the discontent produced by change remains latent at Bournville, less close to the surface than at Longbridge. In some cases, it takes the form of nostalgia or resentment rather than overt aggression:

> 'Planning isn't what it should be. We don't carry stock any more. We're left out in the cold a bit now. You don't know who you're working for.' (Male tradesman)

> 'Management don't care about the workers. All the assets we had are gone.' (Male direct production worker)

> 'It's not such a contented place to work as it used to be. You have to work much, much harder now. . . . Standards have gone down. Change won't happen for the better.' (Part-time female direct production worker)

Finally, three further differences between Longbridge and Bournville should be mentioned. First, the shift towards reliance upon outside contractors as opposed to 'in-house' operations has recently been very marked at Bournville. Longbridge has, of course, long been at the centre of a complex network of outside suppliers. Second, Cadbury retains a sizeable pool of temporary labour on six-monthly contracts which allow the company considerable flexibility in responding to changing production and market requirements. Longbridge has not adopted a similar policy. Third, while the overall thrust of recent change at Austin Rover has been towards centralization of decision making and wage bargaining (one object being to weaken the position of the shop steward), at Cadbury the movement has been towards decentralization so that, for example, the moulded and assortments factories on the Bournville site are currently being managed as separate operations. The contrast between Bournville and Longbridge in this respect can easily be overdrawn, but such variations as exist between them may reflect the differing degrees to which each management feels it can depend upon an informal culture as opposed to a more formal bureaucracy to provide coordination.

Citizens and neighbours

In the industrial sphere the balance of power between capital and labour has recently moved quite strongly in favour of the former. However, workers are also citizens – and voters. The political arena provides an alternative sphere within which influence may be exercised, to some degree at least. Informal networks of kinship and friendship within the two workforces, in so far as they exist, permit shopfloor experience to feed into the subtle and complex process of opinion formation in the political sphere. Other channels of influence on opinion (such as the mass media) exist, as do other relevant arenas of experience (especially the residential neighbourhood). The relevant processes are difficult to untangle but we do know something of kinship and friendship links in south-west Birmingham and also about patterns of voting in recent years. A recent sample of Austin and Cadbury workers from south-west Birmingham, balanced to achieve (as far as was practicable) a reasonable spread between postal districts, age groups and gender, revealed quite high degrees of connectedness through kinship between fellow workers, past and present. Of fifty-nine Longbridge workers interviewed:

- seventeen identified more than one relative as a worker at that factory, past or present;
- another seventeen could name at least one relative in the same category;
- seven had relatives working at Bournville, past or present; and
- only sixteen workers had no kinship connections with either plant.

In the Bournville sample of thirty-seven employees:

- sixteen mentioned more than one relative who had worked there as a Cadbury employee, past or present;
- a further ten referred to a single relative in that context;
- one was related to a Cadbury employee at another factory; and
- a mere ten had no kinship connections with other past or present Cadbury workers.

Although this small sample reveals only relatively week direct kinship linkages *between* the two plants, interviews in local households provide evidence of quite extensive networks mixing kin, neighbours and friends in which Longbridge and Bournville employees, past and present, may both be found. The point of these remarks is that a framework of social relationships extending beyond the factory gates exists within which workplace experience *may* be discussed and evaluated. In view of the prominence of the Longbridge and Bournville factories in the area, *in so far as* workplace-related issues or orientations shape decisions about voting in local and parliamentary elections, then restructuring processes at Austin-Rover and Cadbury-Schweppes *may* have direct or indirect electoral manifestations. After all, Bournville and Longbridge workers have partners (wives, husbands) who also have votes.

Local voting patterns may not be treated as a direct reflection of workplace restructuring at Bournville and Longbridge. However, they are worth examining as one indication of the social and political climate within which restructuring is experienced by citizens of south-west Birmingham who also work at the factories concerned. Most of the locality falls within the constituencies of Northfield (containing Bartley Green, Weoley, Northfield and Longbridge wards) and Selly Oak (into which Kings Norton, Bournville and Selly Oak all fall). Brandwood falls into the constituency of Hall Green.

It is worth looking briefly at parliamentary elections from 1979 to 1987 and municipal elections between 1982 (following boundary changes) and 1987. These include the period of rapid job loss. Following more than a decade of Labour majorities, Northfield constituency fell in 1979 to the Conservative candidate, Jocelyn Cadbury. He took the seat by a margin of 204 on 45 per cent of the vote. Three years later, following his death, Labour regained the seat by less than 300 votes, attracting 36 per cent of the vote in a much reduced overall poll. The following year, at the General Election, the Conservative party captured the seat once again with 43 per cent of the vote, five per cent more than Labour. In the General Election of 1987 the Conservative candidate in Northfield increased his party's share of the vote to over 45 per cent, defeating his Labour and Social Democratic opponents.

If the three general election results (1979, 1983, 1987) are compared, the resulting pattern shows Labour and Conservative neck-and-neck in 1979 (with approximately 45 per cent of the vote each), both main parties losing ground in 1983 (especially Labour), and both regaining ground in 1987 (especially the Conservatives). By 1987, the Tories had regained their 45 per cent share of the vote but Labour won less than 40 per cent of total support.

The pattern just described was disrupted by the 1982 by-election. In that year both main parties lost votes to the Alliance which drew 26 per cent of the vote. However, by 1983 this proportion had fallen back to 19 per cent. It fell still further, to under 16 per cent, in 1987.

In fact, the Alliance polled well in the wards of Northfield constituency during the 1982 municipal elections. Over the next two years its support progressively declined. with the exception of Bartley Green, the main beneficiary was Labour. There were no elections in 1985. By 1986 the Alliance had recovered lost ground to a considerable extent, an advance that was broadly maintained in 1987.

Table 7.2 *Percentage of vote in municipal elections, Northfield constituency (★ = highest poll)*

	1982	1983	1984	1986	1987
Longbridge					
Alliance	26.4	14.6	7.8	19.0	20.5
Labour	42.2★	49.1★	58.2★	51.9★	38.6
Conservative	31.4	35.6	33.5	29.1	40.9★
Weoley					
Alliance	26.3	12.1	9.4	18.3	20.6
Labour	37.9★	43.8	49.8★	49.7★	26.2
Conservative	35.8	44.1★	40.9	36.3	53.2★
Northfield					
Alliance	27.2	11.5	8.9	21.3	20.6
Labour	25.7	35.6	41.9	36.1	26.2
Conservative	47.0★	52.9★	49.2★	42.6★	53.2★
Bartley Green					
Alliance	28.4	23.2	5.8	14.3	13.6
Labour	36.1★	39.8★	46.0	49.9★	34.9
Conserative	35.5	36.1	48.2★	35.0	50.5★

Selly Oak constituency was held by the Tories during the decade to 1974 when it was taken by Tom Litterick for Labour. Anthony Beaumont-Dark regained it for the Conservatives in 1983. Between 1979 and 1983 the Labour vote fell from 49 per cent to 34 per cent while Alliance support increased from 12 per cent to 21 per cent. In 1987 Labour managed to win votes back from the Alliance and brought its own support up to the same level as was achieved in Northfield that same year (just under 40 per cent). Nevertheless, Labour failed to displace the Conservative member whose share of the vote was almost as high as it had been in the previous contest.

In local elections in Selly Oak constituency between 1982 and 1984 there was, as in the case of the Northfield constituency, a swing away from the Alliance (modified by its gain in Selly Oak in 1983) with Labour the chief beneficiary. Once again, support had begun to return to the Alliance by 1986, a minor revival that was maintained in 1987. The same general pattern applies to Brandwood ward.

These results do not permit any straightforward conclusions about the political orientations of local residents working at Austin Rover and Cadbury Schweppes. At first sight it seems that the larger residential clusters of Cadbury workers are likely to be found in Conservative areas (Bournville and Selly Oak) whereas two wards

Table 7.3 *Percentage of vote in municipal elections, Selly Oak constituency and Brandwood ward (★ = highest poll)*

	1982	1983	1984	1986	1987
Kings Norton					
Alliance	18.3	7.2	6.3	11.9	14.0
Labour	44.4★	52.9★	56.3★	53.2★	41.7
Conservative	37.3	40.0	36.5	35.0	43.2★
Brandwood					
Alliance	21.4	10.2	9.7	20.1	19.6
Labour	33.0	40.0	43.4	38.8	29.9
Conservative	44.5★	49.8★	45.9★	41.1★	47.3★
Bournville					
Alliance	25.8	12.4	12.4	22.3	22.3
Labour	29.5	37.0	40.2	36.7	25.9
Conservative	44.7★	50.6★	47.4★	40.9★	49.3
Selly Oak					
Alliance	25.0	12.5	15.5	18.2	19.9
Labour	27.8	36.0	38.1	38.4	31.3
Conservative	42.2★	47.9★	45.5★	43.5★	46.0★

from which Austin workers come in substantial numbers are, in one case, consistently Tory (Northfield) and, in the other case, consistently Labour (Longbridge). Shopfloor employees at both factories were asked about their voting behaviour. Although the responses cannot be translated into statistically precise generalizations about southwest Birmingham employees, it is interesting to note that in both factories the number of workers who had voted Labour in 1983 was approximately equal to the number who had supported non-Labour candidates.

If a strong, positive and consistent political response to restructuring among the local workforce had formed, it would surely have expressed itself at the 1982 Northfield by-election. The previous MP had been a Cadbury. Headlines stressed industrial issues: 'Parties compete in jobs desert' (*Times*, 12 October 1982); 'Northfield could be a test of the unemployment factor' (*Birmingham Post*, 14 October 1982); 'election heats up with BL factor' (*Birmingham Post*, 20 October 1982). In the event, despite considerable local publicity the turnout was only 55 per cent compared to 70 per cent at the previous General Election. Voters withdrew their support from both parties either by staying at home or by favouring the Liberal candidate. Of the contenders for the seat only the Liberal increased his share of the vote. Disillusionment and apathy were noted by commentators reflecting on the campaign (e.g., *Birmingham Post*, 19 October 1982).

The well-known marginality of constituencies such as Selly Oak and Northfield no doubt derives in part from the work of boundary commissioners and from the residents' jaundiced or, perhaps, hard-headed view of politicians. However, this marginality or unpredictability may also be one expression of a complex and subtle political inheritance embodied in the way south-west Birmingham has developed since the late nineteenth century. Broadly speaking, the guiding principle of incoming residents has been the pursuit of improvement. This has sometimes been expressed in terms of the individual's opportunity to get on by securing a well-paid job, giving him or her the prospect of having a comfortable home life. At other times it has been expressed in terms of the community's responsibility to ensure decency and good living for its members, typically through the work of some kind of regulatory agency. Both these interpretations of improvement – the individualistic and the communal – have been present, often closely intermingled.

Each of these ways of pursuing improvement can take varying forms which predispose towards contrary political allegiances and voting patterns. For example, the individualistic self-improver may be (or think like) a small businessman or he or she may adopt an attitude of 'instrumental collectivism' as described in *The Affluent Worker* (Goldthorpe, Lockwood *et al.*, 1969). In other words, support may be given to the Labour Party and specific trade union action solely because of the material benefits gained by the individual. Similarly, improvement expressed through regulation of the community may take varying forms. For example, it may have a traditionally paternalistic form, encouraging acquiescence in an established social hierarchy, or it may be associated with an explicit commitment to radical change. Individuals may switch between these forms of expression, sometimes quite quickly, like the Cadbury shop steward who took voluntary redundancy and set himself up as a small shopkeeper, and the ex-Austin Rover worker who spends a lot of his time, now he is redundant, working and campaigning for the South Birmingham Family Service Unit.

Improvement or survival?

Over the past decade feelings of economic insecurity have increased in south-west Birmingham. However, the work culture among south-west Birmingham residents employed at both Longbridge and Bournville has remained remarkably unchanged. Some feel aggrieved (at Cadbury) that management are failing to maintain standards of consideration set in the past or (at Austin Rover) that the workforce has much less muscle than it used to have. However, they also see capital investments being made and the relative importance of the plants in which they work apparently increasing. In any case, things seem to be no better elsewhere. In many instances, the work is tougher, especially for older people, but at Bournville they can look forward to generous voluntary redundancy arrangements if they are lucky. At Cadbury, the older ones are virtually queueing up to get out.

Those made redundant, especially from Austin Rover, face the danger of considerable social isolation outside the family circle. As a former shop steward at Longbridge commented: 'When people lose their jobs from the factory they seem to lose contact with all their old mates as well'. Redundant car workers join a large but very diversi- fied category of individuals including, for example, unemployed school-leavers and women with children at school entering the job market for the first time or returning to it. The unemployed merge into the larger mass of claimants seeking public support for their families and themselves. To some extent they may become stigma- tized and demoralized.

The unemployed are not organized as a distinctive group within the district. A Restart programme operates locally but the nearest Trades Council Centre for the Unemployed is in Kings Heath, beyond the northern limit of south-west Birmingham. Despite the fact that the working men's clubs of south-west Birmingham do good business, even today, the local Labour movement has not been deeply rooted in strong and well-organized feelings of class solidarity linking workplace and community. It has been driven to a greater degree by the demand that local government should provide ordinary citizens with what was felt to be their rightful inheritance: good homes at reasonable rents and (later) at reasonable prices. The capitalist against which the Brummie has fought, with limited success, is the landlord rather than the industrialist. The 'victories' of the 1950s and 1960s, low municipal rents and the right to buy, helped to provide a relatively widespread material prosperity among working men and women which is now under threat. In fact, the political and social implications of increased unemployment are best treated as part of the broader question of how policy and politics are being shaped in south-west Birmingham in response to the restruc- turing of capital and labour.

South-west Birmingham, whether considered as the home of the 'affluent' car worker or the 'deferential' chocolate worker, was once the kind of place in which researchers sought evidence of social and political convergence. Now it provides ample evidence of polarization at several levels. Birmingham, once bracketed with the 'soft South', has become part of the deprived North. Within south-west Birmingham the outer estates, cut off from established social centres and starved of convenient transport, are showing signs of poverty and distress – violence, ill-health, insanitary conditions – characteristic of the inner city. However, the poor are to be found in every ward and not just on the fringes. They are sometimes cheek-by-jowl with the comfortably off. For example, if you turn up from Bristol Road onto Hole Lane in Bournville ward, you enter a leafy road which passes the opulent Cadbury family home and a private estate of smart executive houses before joining Bunbury Road with its solid middle-class residences. However, if you cross the road and go down the hill you are suddenly in the midst of a series of stark, concrete, high-rise constructions, many until recently in a state of ill-repair. Overbury Road takes you down along the railway tracks through some of the most dismal postwar housing in Birmingham.

Go further south and you are in conservative Northfield which has always aspired to standards of respectability and 'niceness' set in the better parts of Bournville and Selly Oak. Speaking of Selly Oak and the immediately adjacent areas (including Bournville) one of its councillors remarked: 'I represent the community ... there's a massive amount of organizations in the ward and the constituency – residents' associations, all sorts. You have got to be with them. . . . ' Less cosmopolitan than Selly Oak, Bournville is quite self-satisfied and introverted, exuding a sense of parochial completeness. Selly Oak ward has recently undergone quite extensive gentrification. In property terms, it is an 'up and coming area'. Northfield, despite its pleasant terraces and old village centre, has slipped behind from the early 1970s. Between 1971 and 1981, Northfield ward's unemployment rate rose from 4 per cent to 12 per cent. Selly Oak's was marginally higher in 1971 (5 per cent), significantly lower in 1981 (9 per cent). Over the same period, car ownership in Selly Oak ward (which during this period includes much of Bournville) increased from 53 to 64 per cent of all households. In Northfield it rose only from 60 per cent (above Selly Oak) to 64 per cent (equal with Selly Oak).

In fact, the gap between the most comfortably off and the poorest has tended to increase. This is especially the case in outlying areas where diminishing personal incomes and increasingly inadequate public spending (in the face of rising ill-health, deteriorating property and so on) is compounded by the difficulties of paying for public transport to visit faraway shops and other services. Ironically, Longbridge ward in the far south-west has the lowest rate of car ownership in the district. It also has some of the worst housing in the

Plate 7.4   Bournville Village Trust housing.

whole city, especially in Rubery. Typically, the longer-serving car workers have moved out of the poorer houses and maisonettes, on Ley Hill or around Nazareth House for example, and have clambered up the housing hierarchy. This route leads towards older estates such as Frankley and Hawksley or, even better, the nicer parts of Weoley. Many have gone on to the private market. The possible consequences of job loss are visible on the streets they have left behind.

Anxieties and tensions are not exclusive to the poorest estates such as Pool Farm, scene of the recent John Shorthouse shooting (involving an accidental death in a private home during a police operation), or outlying colonies such as Bartley Green, where a social worker was recently murdered. There are frequent complaints of gangs of youths in Frankley (as elsewhere) and even one recent rather lurid report that packs of dogs were savaging old ladies in Weoley. The city council has recently instituted an advice service to arbitrate in quarrels between neighbours.

The spheres of health, housing and education are chronically underfunded. To some extent, increases in central funding of these areas may depend upon south-west Birmingham's continuing political marginality. Two initiatives deserve special mention. One is the attempt by the city council, especially through its Economic Development Unit, to strengthen its links with major local employers, including Austin Rover. The other is the establishment of neighbour-hood offices in several wards throughout the city, including the south-west. These are intended to provide local residents, especially in outlying areas, with more direct access to public services and expert advice. There are neighbourhood offices at, for example, Longbridge and Kings Norton. The work of these offices is sup-plemented by voluntary groups such as the South Birmingham Family Service Unit, also in Kings Norton, which recently (March 1987) won a continuation of public funding despite a prior decision by the social services department to terminate their finance. Sig-nificantly, this local success was the product of an effective campaign for support in the local neighbourhood. Participation by ordinary people may increase if plans to establish a residents board linking the three housing estates in the area are successful.

Conclusions

It is impossible to predict the future shape of the networks and understandings through which people will organize their lives, as far as they are able, within the locality of south-west Birmingham. At present, still buoyed up by the major investments at Longbridge and Bournville in the early 1980s, the district offers at least four models of 'community'. One consists of the networks based on home and kin of the realistic, materialistic, unromantic car workers who, if they engage in politics, tend to use their votes to punish politicians and (if possible) reward themselves. Its underlying spirit might be summed up as 'the wish to be self-satisfied mixed in with resentment in so far as that wish is denied'. A second is the product of the slightly cocooned and inward-looking Bournville environment, whose pro-ducts often seem to feel (without apparent good reason) more secure than they should. The determination to be satisfied may offer an antidote to resentment in this case. A third belongs to the gangs of youths who sometimes steal, burn and terrorize in the worst areas. Resentment without an effective andidote, perhaps? A fourth expresses the commitment of radical community workers to their vision of a world in which residents make decisions for themselves about their own areas and can draw upon the resources and expertise of enthusiastic professionals.

Ironically, while capital increasingly organizes itself within a multinational framework the fashioners of 'community' – deliberate or unintending, malevolent or benign – continue to operate on a very local scale. 'Improvement' has, traditionally, been perceived as the active responsibility of the individual or the community in south-west Birmingham. For decades since before the turn of the century people have gone there to seek it and have expected the locality, so to speak, to deliver the goods. This was plausible when Herbert Austin lived in the Lickey Hills and the paternalistic regime was unchallenged at Bournville. It was still persuasive during the boom years of the 1950s and early 1960s. However, for a long time the major decisions affecting south-west Birmingham's economy, not least its capacity to yield fruits to labour, have been taken far away from the Bristol Road.

Notes

1 This paper is based on research for the South West Birmingham project, one of the locality studies (D04250006) in the ESRC Initiative on the 'Changing Urban and Regional System'. The author of this paper is principal investigator of the South West Birmingham project. Other researchers who contributed to the project at various stages included (in alphabetical order) Susan Barnett, Pat Clark, Susan Ingram, Malcolm Maguire, Michael Maguire, John Newson and Michael Rowlinson. Useful advice in the early stages was received from Ray Loveridge and Richard Whipp. Invaluable cooperation has been given by management and unions at Cadbury-Schweppes and Austin-Rover, local government officers, social workers, local councillors, voluntary workers and others.
2 The terms 'Austin Rover', 'Rover' and 'Austin' are used interchangeably.

References

Church, R. (1979) Herbert Austin. The British Motor Car Industry to 1941.

Dunnett, P. J. S. (1980) The Decline of the British Motor Industry, London: Croom Helm.

Elrington, C. R. and Tillott, P. M. (1964) 'The growth of the city', in W. B. Stephens (ed.), Victoria County History, vol. 7, City of Birmingham and Oxford University Press, pp. 4–25.

Freeman, C. (ed.), (1985) Technological Trends and Employment: vol. 4, Engineering and Vehicles, Aldershot: Gower.

Goldthorpe, J., Lockwood, D. et al., The Affluent Worker in the Class Structure, Cambridge: Cambridge University Press.

Gaffikin, F. and Nickson, A. (1984) Job Crisis and the Multinationals. De-Industrialisation in the West Midlands, Birmingham: Trade Union Resource Centre.

Jones, D. T. (1985) 'Vehicles', in C. Freeman (ed.) Technological Trends and Employment: vol. 4, Engineering and Vehicles, Aldershot: Gower, pp. 128–37.

Smith, B. M. D. (1964) 'Industry and trade, 1880–1960' in W. B. Stephens (ed.), Victoria County History, vol. 7, City of Birmingham and Oxford University Press, pp. 140–208.

Stephens, W. B. (ed.), (1964) *Victoria County History*, vol. 7, City of Birmingham and Oxford University Press.

Sutcliffe, A. and Smith, R. (1974) *History of Birmingham*: vol. 3, *1939–1970*, Oxford: Oxford University Press.

Tomlinson, M., 'Secular architecture', in W. B. Stephens (ed.), *Victoria County History*, vol. 7, City of Birmingham and Oxford University Press, pp. 43–57.

West Midlands County Council Economic Development Unit (WMCC-EDU) (1985) *Employment and Economic Regeneration*.

West Midlands Enterprise Board (WMEB) (1986) *Priorities for Economic Regeneration in the West Midlands*.

West Midland Group (1948) *Conurbation*, London: Architectural Press.

Whipp, R. and Clark, P. (1986) *Innovation and the Auto Industry*, London: Pinter.

Williams, K., Williams, J. and Haslam, C. (1987) *The Breakdown of Austin Rover*, London: Berg.

8

'It's all Falling Apart Here': Coming to Terms with the Future in Teesside

HUW BEYNON[1], RAY HUDSON, JIM LEWIS, DAVID SADLER & ALAN TOWNSEND

Introduction

In the autumn of 1933, as part of his celebrated journey through England, the author J. B. Priestley visited Stockton-on-Tees. From the bridge over the river, he saw:

> the shipyards and slips, the sheds that are beginning to tumble down, the big chimneys that have stopped smoking, the unmoving cranes, and one small ship where once there were dozens. The other men who are standing on this bridge – they have just shuffled up from the Labour Exchange – used to work in those yards and sheds, as riveters and platers and fitters, used to be good men of their hands, but are now, as you can quickly see, not good men of their hands any longer, but are depressed and defeated fellows, sagging and slouching and going grey even in their very cheeks. (Priestley, 1934, p. 343)

What is particularly significant about his account is the comparison which can be made not just with the environmental dereliction and mass unemployment of large parts of Teesside today, but with the causes of this decline. For he was adamant that the demise of shipbuilding at Stockton-on-Tees was one, undesirable consequence of an *international* crisis:

> the world, as we know only too well, entered a period of extreme disorganisation: currencies wobbled, trade declined, tariffs went up, quotas were established, coffee and wheat were burned, fruit was dumped into the sea, and half the bounty of the earth was wasted while half the

[1] Huw Beynon is in the Department of Sociology, University of Manchester; Ray Hudson, Jim Lewis and Alan Townsend in the Department of Geography, University of Durham; and David Sadler in the Department of Geography, St. David's University College, Lampeter.

267

people of the earth drew nearer and nearer to starvation. In this area of rapidly declining trade, very few new ships were wanted; it was not long before good vessels were lying up idle in every estuary. (pp. 344–5)

This meant that

For such a place as Stockton, the game was up. Such new industries as we have had went south. Stockton and the rest, miles from London, and with soaring rates, were useless as centres for new enterprises. They were left to rot. And that would perhaps not have mattered very much, for the bricks and mortar of these towns are not sacred, if it were not for one fact. These places left to rot have people living in them. Some of these people are rotting too. (p. 345)

Is 'the game up' for Teesside today? And what do the 'rules' of this 'game' mean for the ways in which people can and do live their lives? Such questions inform this chapter. It explores how international currents of production and trade (not just in shipbuilding, but in other industries like coal, steel, and chemicals, and more recently even in some service sector activities) have flowed into and out of Teesside. These processes have both shaped and been shaped by the changing social, economic and political character of a changing locality.

Changing times in Teesside

Adventurous, almost intrepid, as Priestley was in his journey through an England ravaged by depression, he did not have it in him to visit Middlesbrough. His comments on this place were short and to the point. It was, he said, 'a product of the new Iron Age'. Its growth was fuelled at first by coal, as the town's harbour despatched the output of the Durham coalfield. Later, the discovery of iron ore in the adjacent Cleveland hills, together with supplies of coking coal from Durham, left Teesside ideally placed to meet the growing demands for iron and steel of a country experiencing industrial revolution. W. E. Gladstone, visiting Middlesbrough in 1862, referred to it as a 'remarkable place, the youngest child of England's enterprise . . . . It is an infant, gentlemen, but it is an infant Hercules' (quoted in Briggs, 1963, p. 241). Sustained growth between the middle and the end of the nineteenth century saw its population rise to over one hundred thousand, where previously there had been only hundreds. And the area came of age, so to speak, with developments in another, newer industry. Across the river at Billingham, Brunner-Mond from 1918, and ICI from 1926 developed a vast chemicals complex, and another company new town sprang up. By 1939 ICI had built 2300 houses as labour migrated into the town. Whereas the steel companies had sought coking coal from the adjoining Durham coalfield, ICI wanted labour power. One ICI manager recalled how:

'There was a lot of very good labour in the Durham coalfields which flocked down to Teesside. They were a very good workforce. We didn't have any union trouble. They were thankful they'd got a job on the surface with fairly reasonable conditions.' (Quoted in Pettigrew, 1985, p. 126)

After 1945 ICI developed a second major chemicals complex at Wilton on the south bank of the river. This expanded under very different conditions of economic development and labour supply. In a national environment of postwar growth, and relatively full employment, the first concern of planning authorities was to ensure at Wilton an adequate supply, not just of raw materials, but most especially of labour. The concerns of the 1930s expressed by Priestley or the flood of immigrants to Billingham must have seemed a million miles away. Certainly this was how it appeared to consultants Pepler and Macfarlane, who reported in 1949 in their Interim Outline plan for the North-east Development Area prepared for the Ministry of Town and Country Planning. 'The development of such a large unit as Wilton on Teesside,' they argued, 'the one area where male labour is in short supply, must entail some influx of population if its labour requirements are to be fully met' (p. 100). Given the high *national* priority accorded to the expanding chemical industry, 'it would be unwise to prejudice the redeployment of Cleveland labour in heavy chemicals . . . by offering it alternative male employment' (p. 76).

Alongside the heightened importance of securing now relatively scarce labour (even if it mean discouraging the introduction of industries which might compete with ICI in its phases of expansion during the 1950s), another feature of the post-1945 period was the increasing attention paid to *planning* for growth. Most significant of such plans was the response to what then seemed a cyclical decline in 1963, typified in Lord Hailsham's report on regional development in the North East (HMSO, 1963). In this, Teesside was to act as a vital part of a regional growth zone capable of generating and attracting new employment to compensate for continuing job losses in the surrounding area, especially the Durham coalfield, reproducing the links established in previous years. Teesside was earmarked for industrial expansion supported by infrastructural investment. In 1965, a land use and transportation strategy was commissioned to provide a framework for this planned expansion. Its optimistic rhetoric captures the spirit of the times:

Plate 8.1 General view of transporter bridge, Offshore Fabricators, Billingham.

Teesside, born in the Industrial Revolution, offers to the second half of the twentieth century both a tremendous challenge and an almost unique opportunity. The challenge lies in the legacy of nineteenth century obsolescence; the opportunity is to make it one of the most productive, efficient and beautiful regions in Britain; a region in which future generations will be able to work in clean and healthy conditions, live in dignity and content and enjoy their leisure in invigorating surroundings. For Teesside already possesses in abundant measure those fundamental characteristics which provide the foundations for a full life. In few places does one find such modern industries, providing for man's economic prosperity, in such close proximity to a beautiful and spacious countryside, which can be the means of satisfying his recreational and spiritual needs. (HMSO, 1969, p. 3)

Teesplan exemplifies a reappraisal of the region's potential which took place in the mid-1960s. It represented a clear expression of contemporary optimism over the prospect of managed expansion in new light industry to compensate for (and indeed far exceed) gentle decline in the traditional industries. The structure of the statutory planning framework meant that in practice, though, there was little opportunity to safeguard the environment. There was no shortage of problems in this direction:

The air along the Tees, full of smoke from its belching factories, forced downwards by cold tidal breezes, must be as badly polluted an anywhere in Britain. It is said that on foggy days, and there are a large number of them, the lightermen steer their barges down the twisting river by the colour of the chemical discharges from the various plants. (Gladstone, 1976, p. 44)

But there was little protest on these matters in Teesside. One housewife had fought a number of environmental battles:

She had resisted a public relations onslaught from ICI, which included exotic alcoholic lunches, and had managed to prove them and other chemical companies in the wrong . . . . When the pollution in the air over Teesside was particularly bad, she would ring up and disturb the Alkali Inspector who lived up in the Cleveland Hills. 'Hullo,' she would say, 'what's the air like up there? It is pretty polluted down here on Teesside.' (Gladstone, 1976, p. 54).

For a period during the late 1960s and early 1970s questions to do with the objectives of planning and planners could be pushed quietly into the background. It looked as if, for once, things were going according to plan on Teesside. The area boomed with new investment, substantially underwritten by the British state. An almost breathless article from the *Sunday Times* in 1976 enthused over this new growth:

If only the spectators could see this. So said Henri Simonet, Vice-President of the European Commission, when he visited Teesside ten days ago. More than a billion pounds is being invested there in steel and chemical plant, nuclear power and oil installations, and the area can fairly claim to be Europe's most dynamic industrial site. But, as Simonet said: 'Nobody in Europe knows about this.' . . . Even now, at a dark moment for the British economy, more than £1,200m is being invested in Teesside, in a series of projects of great boldness, advanced technology and crucial significance for our balance of trade.

Here then was a new prospect for this part of the north-east: high fixed capital investment and outward looking capital meant that the area seemed to have conquered the 'British disease' and turned the corner towards economic prosperity. But even in 1976, at the high point of this 'boom', problems were visible. The clearest was the above average level of unemployment, especially amongst school-leavers.

By 1981 these difficulties had become very apparent, as the *Financial Times* explained:

271

A new beginning with new industries was thought to be the answer. Teesside eventually got that new industry but today it looks back and realises that capital intensive companies are not necessarily the answer. Cleveland has become a model of the new industrial Britain and it still has unemployment problems as serious as almost anywhere in the country.

As Mr John Gillis, the county's planning officer, remarked, 'we have done those things that one is told must be done on a national scale to regenerate British industry.' Clearly, national priorities were not necessarily identical with local needs.

One explanation for the area's economic problems was couched in terms of its lack of diversity. The *Sunday Times* had argued that way in 1976:

Teessiders admit reluctantly that their area lacks the tradition of independent small business that made so many towns in the south prosperous.

The lack of variety was seen as part of the area's heritage, leaving aside any consideration of how and why such a tradition could reproduce itself. Yet, despite this, by all conventional wisdom Cleveland's economic base should have been dynamic:

If ever a place should have boomed, it was Cleveland. It was a test-tube for regional development policies, for the orthodox wisdom of bribing private investors with fat grants. (*Guardian*, 1983a)

In truth, of course, behind the 'tradition' lay the problem of the dominance of big companies like ICI and the British Steel Corporation (BSC) in the local labour market:

If you work for ICI along the River Tees, you can buy your wines on the ICI label, fill up your car at an ICI garage, buy your clothes and suitcases and bedding at the ICI shop . . . . Your local paper will be ICI's Billingham Post, and for senior employees, school fees can be paid by ICI scholarships. 'One of the things that worries me most about this region, even if we have economic recovery, is the psychological dependence of the workforce upon the employer,' says an industrial psychologist with Cleveland County Council. (*Guardian*, 1983b)

One option open to the people of Cleveland in the 1980s differs from the Depression years of the 1930s in scale if not in kind:

This time, the solution will probably be less dramatic, but more practical. They will leave, and emigrate abroad, or move to whatever parts of the country are experiencing a recovery. It has already begun. The population of Cleveland began to decline in the late 1970s, and that is not counting the region's latest export . . . the breadwinners who have gone off to work on the oil rigs and in the Middle East. (*Guardian*, 1983b)

Teesside's record levels of unemployment have increasingly become the focus of outside attention. In 1984 the *Guardian* portrayed Colin Armstrong staring from his third-storey window across to BSC's Lackenby works, which had made him redundant two years previously. Henri Simonet's words – 'if only the speculators could see this' – had a hollow ring to this ex-steelworker:

'If only Mrs Thatcher could see us,' he said. He lives on an estate where 91 per cent of heads of households are unemployed.

By this time the county had the highest rate of unemployment in recession-hit Britain and the echoes of Priestley's *English Journey* were all too loud. Media attention increasingly focused on the 'north–south divide'. Reporters, visibly shaken, expressed surprise at what they saw as stoicism in the face of adversity.

A sense of injustice is widespread. Yet it has produced surprisingly little political radicalism . . . . Stockton is desperate, but it is a quiet desperation. Stockton is angry, but without much hope that things will get better. These are people who are too proud to show how much it hurts. (*Financial Times*, 1985)

Proud or not, the media showed them to the nation. In May 1986 the national BBC TV current affairs programme Panorama reported on 'Hard Times' in Middlesbrough. Presenter Gavin Hewitt spoke of the 'growing number of women for whom separation from their husbands is the only answer to unemployment'; of a 'new jobless society', a 'Giro cheque economy'; and of 'the airport that was intended to entice business to Middlesbrough, which now serves instead the departure of its skilled workers'. This, and the new 'fiddle economy':

In a series of statements we've been told that many of those employed by the contractors are on the dole and not signing off.

A camera crew was directed around the streets of the town to pick-up points used by contractors' vans. The guide showed only the back of his head to the screen, but commented that 'the names you give in are not your own names'. In an equally revealing statement we were told that BSC had indicated that 'what happens between contractors and employees is nothing to do with them'.

The following year the Panorama cameras returned, this time to

Stockton in a commemoration of the death of Harold Macmillan, Lord Stockton. In the House of the Lords in January, 1985, he had commented:

> Sixty-three years ago I went to Stockton-on-Tees to stand for Parliament. The unemployment figure was then 29 per cent. Last November I went to a party there with my friends. The unemployment is 28 per cent. A rather sad end to one's life. (Official Report, House of Lords, 23 January 1985, vol. 459, col. 250).

Echoing the growing political awareness of a North–South divide in contemporary Britain, a strong theme of the programme was the emergence of two nations. The story unfolded from a booming BMW garage at one end of Stockton High Street, to the other end where those who have not, wait outside the town's clubs and discos for those who have, then beat them up as they leave.

The North–South divide in Britain has become an issue of such significance that even the overseas media are interested. In 1987 Canada's leading daily newspaper, the *Globe and Mail*, ran a long, thoughtful story on 'England's great divide'. Among other parts of the UK, its author visited Teesside:

> 'Two nations? It's like the bloody Berlin Wall,' says Gordon, a construction worker in his forties, home on leave in Middlesbrough from the Saudi Arabian oil-fields ... the economy is already losing valuable remittances from workers like Gordon and his friend Dave, as lower oil prices end the Persian Gulf and North Sea oil booms. For both men, their absence in Saudi Arabia has brought divorce and bitterness, and gold Rolex watches studded with diamonds. . . . Dave contemplates his gold Rolex and remarks: 'It's all falling apart here.'

International and national currents of change and the Teesside economy

'The trouble with most local histories of Middlesbrough,' commented one young supervisor of a Community Programme project dealing with just such a topic, 'is that they all seem to ignore the outside world.' Understandable though this is, it is indeed unfortunate, especially given the strong historical ties between Teesside's two dominant manufacturing industries and the national and international markets.

As the iron and steel industry grew and evolved through a process of centralization, the major company to emerge was Dorman Long (see also Boswell, 1983; Hudson and Sadler, 1985). By 1929 it employed 29,000 men and had the capacity to make 1.5 million tonnes of steel annually. It had diversified into engineering and represented an archetypal example of the 'coal combines' which consolidated their control over a variety of branches of production in the north-east during the interwar period. By 1939 it owned eight collieries in County Durham, employing 9000 men to produce 4 million tonnes of coal anually (see Carney et al. 1977; Labour Research Department, 1939). After 1945 these colliery interests were divested to the National Coal Board and the steel business was affected by the short-lived nationalization of 1951.

As part of the British Steel Corporation (a second, more durable exercise in nationalization) from 1967 onwards the future of steel production in Teesside seemed secure, especially with the announcement of BSC's Ten Year Development Strategy in 1973 (HMSO, 1973). With world steel demand forecast to grow at 4–5 per cent annually up to 1980, this proposed massive investment in a new steel complex at Teesside with an ultimate annual capacity of 12 million tonnes. Forecast growth in steel demand failed to materialize, though, and world steel output actually fell for the first time since 1945. A halt was called to the 1973 expansion programme, effectively from 1976, formally announced in 1978 (HMSO, 1978). Instead of expansion, employment in steel production on Teesside has been slashed from 29,000 in 1971 to just 7000. The truncated development of the BSC South Teesside works has left it in a unique position, the only major steel complex using the basic oxygen route dependent solely upon one blast furnace. A second, projected blast furnace lies rusting in an adjacent field, never constructed. The single blast furnace now uses predominantly imported coking coal, with one colliery after another closing on the south-east Durham coalfield in consequence (see Beynon et al., 1986a). Dependence upon one blast furnace, subject to periodic interruptions of production to replace the refractory brick lining, leaves the works vulnerable to continued plans to cut capacity at BSC. In 1982 and 1983 real fears of complete closure were expressed locally (see for example Cleveland County Council, 1983; Hudson and Sadler, 1984), only (temporarily?) quelled by an elaborate plan to cope with the relining of the Redcar blast furnace during 1986. Additionally, BSC cut back rolling mill capacity, leaving the works deficient in profitable, finished product activities. Its main strength is in semi-finished steel products, open to cost-effective competition from Third World steel industries. BSC also divested itself of the construction and fabrication activities acquired via Dorman Long's earlier diversification, including the sale of offshire construction interests to the Trafalgar House group (see Sadler, 1986). Since 1981 BSC has been subcontracting out parts of its

Plate 8.2 Clay Lane blast furnaces and South Bank housing.

operations, as it has at the rest of the 'big five' coastal complexes (see Fevre, 1986). The scale of this is lost upon local union officials. One AUEW convener has expressed astonishment at the number of subcontractors who 'materialized' out of the works during a fire practice drill.

Similar processes can be observed in Teesside's second major manufacturing employer, the chemicals giant ICI (see Beynon et al., 1986b). The very name – Imperial Chemical Industries – conjures up images of an industry developing from the outset to serve an international market. As the protected outlets of the old Empire disappeared in the postwar period, the company expanded instead into other international markets. The interwar growth of a major inorganic chemicals complex at Billingham, and postwar development of the organic complex at Wilton, played important roles in this global strategy (see Pettigrew, 1985; Reader 1970, 1975). ICI has invested heavily in the most up-to-date technologies to meet an overall growth in demand for chemicals and to protect its position against competition. From 1962 onwards, for example, major replacement investment took place in ammonia production at Billingham, replacing coal and coke as a basic feedstock first with naphtha, then later with natural gas. Larger production units and cheaper feedstock dramatically reduced costs; in addition, labour requirements were slashed and several thousand jobs were shed. Continuing employment growth in new plant across the river at Wilton, though, meant that in this period most of the job losses could be accommodated through transfers.

In the 1970s, as demand for bulk chemicals slumped, ICI found itself facing a new, intensified set of international competitive pressures. Overcapacity in ethylene production, the basic building block of most plastics such as those produced at the Wilton site, was increasingly in evidence throughout Europe, the USA and Japan. In response, ICI initiated a series of plant closures at Wilton, especially after 1980 when the company recorded its first-ever net loss. In a letter to all MPs in 1982, the company even threatened to close the entire Wilton site, claiming it was suffering unfairly from the tax concessions granted to its UK competitors over their supplies of ethane, in contrast to the tax arrangements for its naphtha-based plant at Wilton. Four years later ICI won a prolonged court battle over this issue and, with crude oil prices tumbling, petrochemical profits increased again. Nonetheless, intensive competitive pressures remain from companies and countries with access to still cheaper feedstocks, most especially in the Middle East. Billingham's main product, agricultural fertilizers, is also a market area under great strain in the UK, with demand falling and a strong competitor emerging since 1982 in the Norwegian conglomerate Norsk Hydro.

In response to such pressures in the UK market, ICI has increasingly located production overseas. The Wilton works, for example, now has a parallel production facility at Wilhelmshaven in West Germany. ICI employment in the UK has fallen from three quarters of global company employment at the start of the 1970s to less than a half. On Teesside this has meant a reduction from 31,500 jobs at ICI in 1965 to 14,500 by 1985. In addition, priority has shifted away from the so-called 'bulk' or commodity chemicals of plastics, petrochemicals and fibres, and into speciality chemicals where the emphasis is on high value-added, low-volume production. Employment in the manufacture of these new chemicals is relatively low and it is a high-risk business. The most significant such investment on Teesside, the 'Pruteen' plant, was opened in 1980 but has only operated intermittently – and when it does, it employs a handful of people to produce just 150 tonnes of artificial protein daily.

ICI's labour policies have also evolved in a sophisticated fashion over the years. From the earliest days, trade union activity was incorporated into ICI consultative councils as a deliberate technique of the company's personnel department. This was especially marked at Billingham in the interwar period, where the expanding company town was largely serviced by the ICI Works Council, not the town council. Wilton grew under very different conditions of labour market supply and was characterized, for example, by more marked resistance to ICI's major work measurement scheme, the 1969 Weekly Staff Agreement (see Beynon et al., 1986b; Roeber, 1975). Wilton workers were also more actively involved in a loose ICI combine committee, eschewing the more collaborative form of trade unionist.

The use of contractors' labour has escalated significantly, especially for maintenance. In 1984 ICI proposed that process workers should do some maintenance work. Whilst this met with strong union opposition, a more portentous move was announced in 1986 and implemented over the heads of trade unions the following year – the hiving off from ICI of the commodity chemicals side to a new subsidiary company, ICI Chemicals and Polymers, in which redundancy and demarcation agreements are completely open to renegotiation. The future prospects for ICI employment on Teesside are grim. Large-scale, continuing job losses for at least the next five years have already been publicly forecast by senior ICI management.

ICI brings a high level of political awareness to this process of managing decline in the UK. 'Resettlement' schemes were established at Wilton from 1981 to encourage workers to find alternative employment or (mostly) retire. The promotional agency Saatchi and Saatchi has been appointed with a £10 million budget to highlight ICI's strengths. In 1987 a science park was established alongside ICI Billingham, with the intention of attracting alternative jobs into small factory units (and deflecting criticism of the company's labour shedding). Throughout its history, and especially recently, ICI has shown itself keenly aware of the need for large employers to maintain a political presence both locally and nationally.

Like ICI, other sectors of the Teesside economy display both continuity and change in decline. Not all sectors of employment in the area are in such drastic decline, of course. *Teesplan* was strident in its insistence that there was a need to diversify the employment structure through expansion as a sub-regional service centre (HMSO, 1969). To some extent this happened, though on nothing like the scale necessary to mop up job losses elsewhere in the local economy. Many of the spaces reserved for office development now stand as vast 'temporary' car parks on the fringes of Middlesbrough town centre, a silent testimony to unfulfilled expectations. A net decline of 20,000 manufacturing jobs in the period 1971–1981 far exceeded a net gain of 4000 service jobs – with most of this latter increase as female part-time employment. Full-time service sector employment actually fell. By 1981 chemicals and steel still accounted for 60 per cent of all manufacturing jobs but over a half of Teesside's jobs were in the

Plate 8.3   Barclaycard offices.

service sector. The rise of the service sector, if sometimes exaggerated and somewhat more complex than it is often presented, is nevertheless significant, especially in bringing large numbers of (married) women back into the labour market or into employment for the first time.

Much of this service sector growth was in financial services such as banking and insurance, which increased employment by 80 per cent between 1971 and 1981 (see Lewis, 1987). This has been vulnerable to technological changes, particularly the increased use of networked computers. Many insurance and finance companies have recently reorganized their operations within the north-east, typically concentrating on Newcastle and closing offices in Middlesbrough. As credit boomed via plastic money, a significant component of financial services employment in Teesside has arisen in Barclaycard, which opened offices at Middlesbrough and Stockton in 1973 and 1974 respectively. In 1985 these employed 700 people, 85 per cent of them women. The main tasks are processing of remittances and sales vouchers – entailing the continuous use of a VDU – and handling customer enquiries.

Service employment in the public sector is typified by the National Health Service. Here too the workforce is predominantly female, but the private sector has not dominated development as it has in financial

Plate 8.4   Remains of Britannic Steelworks.

services. There are no private hospitals in the area and there is a strong commitment to the public provision of health care through the NHS. The industrial and urban legacy of the area is apparent in its relatively large numbers of hospitals and hospital beds, with a standardized mortality rate 22 per cent above the national average (see Townsend, A., 1986; Townsend, P. *et al.*, 1986). In certain areas close to the river, mortality rates are even higher, one impact of long-term environmental pollution. The main planning issue for the South Tees Health Authority concerns the fragmentation among a large number of aging hospitals. Recent years have seen centralization of investment on the new South Cleveland hospital in Middlesbrough and a continuing tension over patient access to facilities in east Cleveland. Centralization in south Cleveland has not been opposed by the trade unions but they have been strong opponents of a more recent development, the planned contracting out, via privatization, of tasks such as catering, cleaning and domestic work.

These changes in steel and chemical production, and in financial and health service provision, represent a series of portraits of the dominant characteristics of change in the economic structure of Teesside. The two major manufacturing industries have acted in a changing and increasingly competitive national and international environment to attempt to secure continuing production, with a considerable degree of state support in the form of investment subsidies and, in the case of steel nationalization. Regional Development Grant payments and the debt financing of BSC served to underpin investment in new technologies but did not, indeed could not, generate employment gains as in the initial period of absolute expansion of production. Both steel and chemicals industries currently rest on a precarious toehold in Teesside, subject to overseas competition in export markets and import penetration of the UK market. While service sector activities have become, almost by default, of greater significance than manufacturing in terms of the number of jobs, these are often of a qualitatively different character. Service sector growth has been dominated by unskilled and poorly paid part-time female employment, vulnerable to renewed technological change in the case of financial services, and to government-imposed financing limits in the case of the health service.

Such developments in these four very different industries mesh together in the character of the Teesside labour market. The dominant feature of this is the shortage of jobs. Cleveland County now has the highest rate of unemployment in Britain. In such a climate, changes in labour practices including the use of subcontractors or the spread of 'flexibility', both in and out of work and within work, are more easily imposed. Pressures in this direction, as we have seen, are evident in both manufacturing and service sectors, and indeed are increasingly apparent nationally (see NEDO, 1986). On Teesside there is a further emergent trend, one rooted in the earlier prevalence of skilled yet temporary employment in building the new chemicals and steel plants – that of migration of skilled construction workers overseas on lucrative, short-term contract work, to areas such as the Middle East. The irony could not be sharper. As Teesside's industries stagnate or die, Teesside's workers – at least those with the necessary skills acquired in an earlier era – are forced to find employment in precisely those countries where competing industries are emerging. Britain's role within an evolving international division of labour could not be clearer.

Politics and labour in Teesside

Within Teesside these labour market changes, both historic and recent, have recognizable implications. They illustrate how national plans and priorities have varying and sometimes unintended consequences in different localities as state intervention is unable to abolish the contradictions of capitalist production. At different periods, different processes have characterized the development of

Teesside as a locality. These have been given order through political apparatuses. Teesside represents perhaps the clearest illustration in the UK of the linkages between capital and labour exemplified by ICI's paternalism, and solidified as a political practice by Labourism. This is most apparent in the provision of housing.

First seemingly endless growth, then more recently economic stagnation, depended upon the construction and reproduction of a series of physical and social relationships between industry and the built environment of roads, housing, schools, shops and other means of subsistence. Much of the early expansion of steel and chemicals production depended upon the construction of successive company towns, first in the early Ironmasters district of nineteenth century development in the heart of Middlesbrough, then later at Billingham and in new estates for steel workers such as Dormanstown. Much of this growth was fuelled by substantial inward migration from other parts of the north-east, or even further afield. At times of labour shortage, such as immediately after 1945, the state also became heavily and visibly implicated in accommodating the labour force required by the expansion of manufacturing industry through the provision of council housing.

Such concerns intertwined with the plans of local authorities to clear a vast amount of run-down housing in this early postwar period. Middlesbrough County Borough Council commissioned a detailed survey partly to cover this task (see Lock, 1946). It reported that 'on strictly utilitarian grounds' the old Ironmasters district, the St Hilda's ward, should be cleared completely of its stock of slums (owned principally by a dozen private property trusts and agents). It also recommended that if the town grew strongly through immigration it might consider establishing a satellite community. Both were long-term projects – St Hilda's now has a new stock of partnership trust housing built jointly by Middlesbrough Borough Council and Yuill's, a firm of private builders, whilst substantial growth is currently taking place in the Coulby Newham estate south of the main built-up area. The role of the state in mediating labour and housing markets is clearly apparent in the new conditions under which such policies evolved after 1945.

There is a historical consensus within the area that the way forward is via compromise between capital and labour. This is largely grounded in an argument that what is good for Teesside is good for both employer and employed. In a sense, there exists a Teesside-based coalition of interests, although this is being placed under increasing strain by both the evidence of past failures and the increasing external control of the 'local economy'. Much of this, to be sure, is rooted in the early paternalism of the steel and chemicals employers; but whereas from other similar industries such as coal mining there grew a relatively strong, independent trade union organization, the same cannot be said for Teesside.

From 1945 onwards the Teesside Industrial Development Board (TIDB) was an important cross-party vehicle, incorporating representatives of most local authorities, the local chambers of trade and commerce and trade union organizations such as the Iron and Steel Trades Confederation, the Electrical Trades Union, the Amalgamated Engineering Union and the Confederation of Shipbuilding and Engineering Unions. A revealing insight into its character (and that of the area) was presented when the need for a reorganization of the North-east Industrial and Development Association (NEIDA) was agreed in 1960, after the failure of the region to 'win' any of the 1950s round of new car plant ventures. The TIDB was concerned to secure three organizations – centred on the Tyne, Wear and Tees – in a federal structure, leaving the Tees considerable autonomy, most especially to cooperate with local industrialists. Such a proposal was not popular with NEIDA, which adopted a new constitution in 1961, forming the North-east Development Council (NEDC). The TIDB reluctantly affiliated to NEDC but some Teesside-based local authorities, including Middlesbrough, did not.

In this period the coalition of interests typified by the TIDB could be held together around a consensus on the need for and possibility of growth through modernization of the sub-regional economy. The economic downturn of 1962, to which the Hailsham white paper on the north-east was one reaction (HMSO, 1963), represented only a slight hiccup, and if anything the proposed solution served further to coalesce interests and secure agreement. TIDB president, Mr Robson, assured the Board's Annual General Meeting in 1965 that there was 'concrete evidence of great progress on Teesside'. The following year Dorman Long contacted the Ford Motor Company indicating the suitability of the Ironmasters district for a projected new car factory. Teesside was viewed, at least locally, as a prospective candidate for any potential investment project.

With the collapse of such growth policies in the 1970s and their failure to secure *either* employment stability *or* a sound industrial structure, their fragile logic was increasingly apparent and the coalition built around them has come under considerable strain. This was epitomized by James Tinn, Labour MP for Redcar, speaking in 1979:

> I remind the House that in the last decade no less than 10,000 jobs [in steel] were lost on Teesside. In only one closure was there a massive and well-organised protest, and that was not in my constituency but across the river. The Teesside workers recognised that modernisation was inevitable

and that a price had to be paid for it. They paid the price before they got the new works. (*Hansard*, vol. 973, 7 November 1979, col. 490).

His concern was with the possibility of further works closures, arguing that this would be unfair on a workforce which had directly cooperated with a decimation of the steel industry for the promised 'new works'. That this was already at risk from competition even as it was being built had been quietly pushed into the background.

The growth conditions of the 1960s represented one expression of a national consensus, evident in the National Plan (HMSO, 1965). The failure of such policies is also of considerable national significance, apparent in emerging policy responses. The currently dominant Thatcherite philosophy is that it was a mistake for governments to seek to plan industrial development, and that policy should confine itself to ensuring the national and local preconditions for private investment, which can and should take care of itself. Such is the contemporary political climate that this represents the best which even relatively progressive (but ratecapped) local authorities such as Middlesbrough Borough Council feel they can offer, and the Iron-masters district became an Enterprise Zone in 1983. More recently, designation of Teesside as one of the four Urban Development Corporations represents a central attempt to take control over a range of functions away from local government altogether. Long-serving local councillor Mrs Maureen Taylor is in 'no doubt whatsoever that the UDCs represent an attempt to destroy local government'. Local politicians are being placed in a position where the agenda is constrained by a perceived need to argue precisely for local economic initiatives, even though their impact is recognizably limited.

The alternative Conservative response is to argue that the earlier 1960s consensus was correct; that governments could and should create jobs and reduce unemployment through direct investment support to private industry. This concern was forcefully expressed by Edward Heath MP on a visit to the north-east. In 1963, as Lord Hailsham prepared his white paper, Heath had been Secretary of State for Trade and Industry. Later, as Prime Minister, he was responsible for its implementation. In 1985 he commented that 'there are lessons to be learned. The first is that a strategy can be developed to deal with the problems of economic decline, the second is that such a strategy can be successfully implemented.' (Heath, 1985)

Whilst some objectives of the strategy in terms of new infra-structural provision, and the restructuring of major capitals engaged in steel and chemicals production, might have been successfully achieved, the employment aims clearly were not. And herein lies the greatest challenge for both industrial policy and Labourism in regions such as the north-east. This has been most widely recognized in the adjacent area of the Durham coalfield, intimately linked with Teesside, physically and socially, for over 150 years. Durham miners on strike in 1984 and 1985 often contrasted their dispute with those of 1972 and 1974. In the earlier disputes the flow of coal from Durham to the steelworks was stopped and the works picketed so effectively that no alternative supplies reached them. Durham coal was also then transported to power stations on Teesside. These are now closed and their capacity taken up (theoretically at least) by the nuclear power station at Seaton Carew, near Hartlepool. British Steel now imports almost all its coking coal through a harbour designed originally from iron ore imports. Ideas of a Triple Alliance between coal, steel and rail workers proved fragile in 1984/85. In May 1984 Alan Cummings, NUM lodge secretary at Easington Colliery, commented that 'BSC has been hoodwinking the unions and we want coal imports into Redcar reduced dramatically', whilst Mr Leadley, Middlesbrough-based general secretary of the National Union of Blastfurnacemen, concerned for the future of the steel complex, responded that 'we will be looking to hold on to what we have got' (*Middlesbrough Evening Gazette*, 1984). This intransigence led to an increasingly bitter breakdown of whatever alliances had existed between Teesside steel workers and Durham coal miners.

These shifts in the political culture of the coalfield and the Teesside area are deeply significant. They epitomize the ways in which a coherent local economy in this part of the north-east region has been placed under increasing strain by international processes of change, mediated by national and local policies and politics. In different parts of Teesside, changes in labour and housing markets have historically come together to produce varied conditions of economic and social life. Underlying this variation has been a Teesside-based consensus on the need for growth through modernization, held together only so long as that growth could be maintained, and linked to the adjacent coalfield through a series of transitory transfers of labour and commodities. Today these ties are seemingly broken and the consensus is being torn apart. In the process what are raised are some fundamental political questions to do with the role of state policies, the purpose of production and the conditions of everyday life. These changes are being experienced and expressed by people on Teesside.

Experiences of life in Teesside

One group of people with an insight into politics and labourism in Teesside are trade unionists in the dominant manufacturing industries of steel and chemicals. Such views are often formed through long experience of negotiating with ICI and BSC. It is instructive to contrast their impressions of the respective management styles. One local AUEW official, Bill Purvis, reflected on the character of ICI in this way:

'ICI kill you with kindness. They consult so far ahead, it doesn't register. You don't think what's going to happen in twelve months' time – you might be dead by then. Then before we know where we are, twelve months later, ICI say there's been no response so they assume it's all right to carry on. It's like wading through cotton wool. At times I've sat down at night and thought, "I've been conned."'

This comment on ICI's ability to think long-term and strategically in contrast to trade unions' reactive character was echoed by one TGWU shop steward from Billingham, analyzing the formation of a new subsidiary company. 'They've been planning this for at least five years', he said. In many senses ICI is regarded in a different light to BSC. An AUEW convenor at BSC, who had worked previously at ICI, compared them in this way: 'You're talking about professionals versus amateurs, in every respect.'

Whilst there is considerable agreement that ICI's formal management style displays great professionalism in its handling of labour issues, a fair degree of control and coercion is exerted informally by senior BSC management. One recent works manager always met informally with selected trade union officials, every Friday, 'to tell us what was going on', as the same convener put it. Another AUEW convenor, Ron Agar, went into more detail as to how these meetings were arranged:

'Consultation with management got so bad in 1981/1982 that we phoned MacGregor in London. He was at lunch but he phoned us back at two-thirty. He said, consultation must take place – there's two directors you've got there, talk to them. There followed meetings with local management, once a week, every week. They were very open – he would tell us things we hadn't to disclose outside the room. He would tell us about investment when it was a thought in his own mind, before it even got to the planning stage.'

Such meetings are of course as much a comment on the character of local trade union leaders and of trade unionism within the steel industry as of management style. To an extent today, this is conditioned by fear of total closure, expressed in both steel and chemicals industries. At BSC, as an alternative to shutdown, possible privatization is far from unpalatable. In this the need to make a profit is promoted by union officials. 'They say you've got to make profits and I go along with that – without profit there's no money for investment,' commented Ron Agar. He went on to speculate about the future:

> 'I can see within the next three years we'll be ripe for privatization, if the political climate stays the same. Then we can compete better with the Japanese and the Germans on productivity.'

His main reason for wishing to remain part of BSC was that employment conditions – pensions, sick leave and pay – were preferable to what he expected from the private sector. The significance of such a position over the prospect of privatization is not lost upon some union officials and members. One ISTC shop steward has seen the extent of changes entailed by privatization in the canteens:

> 'The biggest change is, you can't do the job properly. They know the standards are well below par, they're putting us under pressure to up the standards, but we can't do it, we've not got the time. There's not enough staff to cope. It's better to work for BSC than for a contractor.'

She went on to compare BSC management with an earlier era:

> 'It's all wrong, I mean Bolckow and Vaughan (Victorian ironmasters) would turn over in their graves. There are people in that place today, who just don't care what's going to happen to Teesside.

Whilst, as this woman recognized, the earlier paternalist links between manufacturing employers and their workers in the area have been broken, there clearly remains a legacy of compromise between senior management and trade union leaders. The implications of this basis for agreement in an evolving labour market, and for political life, are increasingly more complex.

High unemployment does not directly affect all households to the same degree or in the same way, but the depressed labour market does increasingly have a general effect on those in work or seeking work. One household affected in this way is that of the Cowdreys, a couple in their thirties who live in Acklam with their two young sons. Martin held a variety of jobs, later joining British Steel. In 1982 his part of the works went out to private contractors and he was laid off. It took him two years to find another job. 'I thought I'd probably get another job straight away, but it wasn't to be.' Now he works as a driver on the buses. After two years back at work, he was still struggling to pay off the bills, even though he had invested his redundancy payment wisely. His wife's earnings helped as well, but without the redundancy money 'there's no way we could have stayed here with the mortgage – we'd have had to think about moving into the town'. He was relatively lucky – he'd just heard from one of the men laid off from British Steel at the same time, who still hadn't found a job. Carol worked in a number of office jobs, most recently at Barclaycard. She regularly works overtime at weekends, to earn a bit extra for the holidays. She can't see their sons getting a job locally.

'A few years ago you could pick and choose your own jobs, but if you've got a job now you're better off sticking with it. My first job I left because I didn't like the bloke I worked with, but now I wouldn't do that.'

In many ways the Morecambes are a similar household. Richard and Lucy have been buying their council house at Coulby Newham for several years. Richard works shifts at ICI so that they have to negotiate over who has the car because the estate is quite a way out of town. Richard started his early working life in the steelworks before moving around the industrial plant in the area. 'It's a funny thing, that, I always stayed in one job for about two years,' he remarked. In the early 1970s they thought about moving away, but then he got a job with ICI. Some weeks before we first met he was told that his part of the works was soon to be closed, and he was in the process of evaluating his options once he was laid off. He's convinced the chemicals industry is dying on Teesside. 'When I first went to Wilton it took you half an hour to get into the site because of all the cars queuing to get in. Now you drive from one end to the other in a matter of minutes.' At night, the evidence of decline is even more apparent. 'There's plant that should have lights on, and there's nothing. They've closed down – vast open spaces in the middle of the works.' If Lucy didn't work, Richard would work overtime. As it is they've got by, although they're obviously concerned for the future. Lucy said she'll encourage their two young daughters to travel and see the world – 'because I don't think there's going to be anything for them here'.

Even at service sector employers like Barclaycard, the pressures in work are increasing. Irene and Tom Oldham live in Acklam with their two daughters. Irene wants to work and almost always has. 'I have to work, it would drive me mad not to work, just being a housewife. I want something that I can do, a bit of independence. But we were all a lot happier three or four years ago when there was no pressure on.' She recognizes something else: that at Barclaycard, 'few women have husbands who are unemployed. It seems people who work in places like that don't come across unemployment so much.'

There's different sets of people. Tom has always maintained that if he lost his job he'd find another one.' This he has done successfully on numerous occasions in the past, and both their daughters now also have office jobs.

In their various ways, these households are illustrative of a range of people in Teesside who have evolved ways of coping with the exigencies of its changing labour market. What they indicate too is a growing polarization of society within Teesside between those who are given, or take, the opportunity to adjust to a lifetime of insecure, shifting or intensified employment, and those who do not even get that chance (see also Foord et al, 1985, 1987). Paula, an unemployed factory worker, expressed her frustration at this growing divide within the area:

'People you meet on the street say 'Aren't you working yet?' but they don't understand what it's like being unemployed. I think to myself, they read the papers, they see the news, but they don't really know what's going on.

In a work-oriented society, unemployment frequently brings domestic tensions. One fifteen-year-old recalled how his father was out of work for more than a year.

'He got very depressed; he took up golf, but couldn't stick it. He just started getting depressed and niggly. You couldn't blame him really. You could understand. He was used not so much to a lot of excitement, but pressure. I think my parents nearly did split up at that time.'

Becoming unemployed has been an everyday experience for some time now in Teesside. It is an area which has experienced the effects of recession before, in the 1920s and 1930s (see Nicholas, 1986). One man laid off from the engineering industry put this process into words. He described it as 'just a small item on page seven – you know, the industrial news – and you're out of a job'. Finding another one, or a first one, is not so easy. In the Middlesbrough Job Centre a twenty-three year old butcher, unemployed for two years, commented:

'I've lost count how many jobs I've applied for. I'd say about fifty. Now I've got to force myself to go and look at the cards on the board. A lot of people have given up – just sit at home saying there's no jobs so there's no point looking.'

One reason for giving up lies in a perception of what employers are looking for. Mary, an unemployed catering worker, described one job she had applied for:

'At 53 I was too old. The woman behind me in the queue was too young at 20! If you want a job these days you have to be between 25 and 35. If you're old you've no choice. But there doesn't seem to be any jobs for younger ones either.'

Increasingly, finding a job is as much about who you know as what you know. There is widespread, tacit acknowledgement of the significance of informal networks in finding work. 'Though a friend of a friend' is frequently heard in this context. Sometimes these connections extend into the greyer areas of the labour market, although there is more than a suspicion that the extent of the so-called 'black economy' is overstated and the benefits to potential (often waged) consumers under-emphasized, as living 'on the fiddle' is fraught with dangers. One unemployed man put it this way:

'There's a lot of people on the fiddle – but then again a lot are jealous of the money they earn. There's a lot of anonymous phone calls to the DHSS. I wouldn't tell on anybody myself, but I wouldn't take part either for fear of being caught.'

Moving away in search of work is one escape sometimes seen as an inevitable process. A local businessman, for example, felt that 'nobody goes from the South to the North, it's against the natural routes'. Not that there is anything necessarily 'natural' about a journey in the other direction. It has its complications, and can be very traumatic. A school-leaver explained how his father worked in the Middle East on contract work.

'A year ago he came back from Oman. It's expensive to take the rest of the family over there so he has to go out for three months at a time and when he comes back, he's only back for a week and a half or something, so you don't see much of him. You get used to it – he's always away. Now, he's not away, he's working in Watford, down south. He works weekdays and he's back at the weekend.

As in many parts of Britain, party politics plays little formal role in everyday life, for the waged and the unemployed alike. Part of the reason for this lies in a cynicism, a disbelief in the power, or even intention, of politicians to deliver. This was neatly expressed in the following unprompted dialogue between Tom and Albert, two council workers on the verge of retirement.

Albert: The only way is to get rid of Mrs Thatcher.
Tom: But look what happened when Labour was in – everything was on strike and got paid off after six or eight weeks. At least she's said no, you can go on strike but we're not paying you.

Albert: But at least the money she's spent on strikes would have been better paying people to work.

Tom: It's gone down from one government to another. This one's no better than the last, it's all the same.

Albert: You look back though, history shows the Conservatives can run the country better with more people unemployed.

Tom: Yes, it's the policy of this government, now, there's no doubt about that.

Albert: But if Labour got into power tomorrow, they couldn't do a great deal about it.

Tom: Yes, no doubt.

Given this acute perception of the limits to what government policies can in practice deliver, it seems to make increasing sense for households and individuals to adopt their own (in the final analysis, competitive) strategies of 'getting by'. Yet ultimately, of course, the diverse problems of the area have been structural roots.

One reaction to Teesside's new situation is to seek to extend the old; to argue that because the area is experiencing today conditions which the rest of the UK will ultimately have to experience, Teesside should be a testing ground for policies to cope with problems of long-term mass unemployment. One senior local government officer with Cleveland County Council, Reg Fox, put it like this:

'The problem we've got is that we've been the forerunner of changes in manufacturing, and I don't think society generally and the national political scene has caught up with the fact that this is going to be happening in the rest of manufacturing, and until the country comes to terms with that, we'll not have the resources to tackle our problems.'

To which another, John Gillis, responded:

'The sad thing is, for at least the last ten years we've been telling people this. We must cope with these people who are unemployed. What are we going to do with them?'

This expression of despair partly masks an unspoken recognition that Teesside as a place both exemplifies national patterns of change and is increasingly marginal to the main currents of the world economy. There is nothing new about Teesside's susceptibility to international change – J. B. Priestley recognized it in the Stockton shipyards in 1933. There are limits to what a capitalist nation state can and will do in an international competitive economy. One way of coping with this is seen as a reversion to the local: to devise, develop and implement political strategies at a local level knowing they are

291

constrained both in terms of resources and the agenda which they address.

There is a deepening recognition within the area that local economic development initiatives, heavily constrained as they are by a hostile central government policy, are of limited effectiveness. Cleveland County Council's leader Bryan Hanson weighs the balance in this way: 'One year's activity by the county council in the job creation field can be wiped out by one plant closure.' The expansion of make-work schemes by central government, through the Manpower Services Commission, exemplifies national policy towards the unemployed in places like Teesside (see Benn and Fairley, 1986). Whilst it is a significant transfer of financial resources in aggregate (though not to each recipient), the area lacks power to control this expenditure. Yet the unemployed remain a significant, if changing population in Teesside, and the county council has therefore formulated its own unemployment strategy (see Smith, 1986). It is based on the premiss that 'policies should not be devised *for* or *at* the unemployed, but *with* the unemployed' (Cleveland County Council, 1987, p. 2). The evolution of this initiative was described by Councillor David Walsh as part of a growing recognition that we can no longer rely on government intervention, that with a government which felt the region had played out its historic role, we have to take things into our own hands'. Yet, without adequate finance, it is also clearly accepted that the proposals will not solve the problem of unemployment, or relieve the misery it causes (Cleveland County Council, 1987, p. 3).

In this sense, then, the area is undergoing a potentially dramatic political transition. David Walsh perceives:

'a huge generation gap in the Council Chamber, between people in their twenties, many unemployed, and those in their sixties, who grew up in a different era. People who are in their forties now grew up in a different period again, and didn't get involved politically.'

There are emergent signs of these political changes elsewhere in the locality, too: the retirement of James Tinn, long-serving MP for Redcar, and the replacement of Arthur Seed, leader of Langbaurgh District Council on the south bank of the river, to name just two locally well-publicized examples. Out with the old guard, in with the new; but what they make of the area is a matter for the future. The outstanding challenge is to inform the (admittedly hard) choices which need to be made if people in places such as Teesside are to avoid hopeless future material deprivation and grim despair.

Some concluding comments

The main features of Teesside's recent economic history are clear and easily understood – a transition from boom town to slump city; an area now of high unemployment where capital intensive investment created few new jobs, where service sector growth was unable to compensate for manufacturing decline, leaving migration in search of work the only realistic option for a substantial minority. Yet the story is also a complex and intricate one, because its plot was written by so many different and competing interests. Diversity in decline is a marked characteristic of Teesside, and other areas of the north, today. In this tension lies a source of polarization and fragmentation.

The evolution of Teesside has been a social process, entailing political decisions. Its current situation poses acute questions, also of a political nature, which are significant both nationally and locally. Different phases of Teesside's development have coincided with different dominant conceptions of the appropriate relationship between national and local government. In the boom years of the 1960s and early 1970s, the area was a highly attractive location for some forms of capital investment in manufacturing. Local authorities in that era were engaged not so much in planning for growth, as responding to growth. Planning was orchestrated nationally whilst, as Cleveland County Council's leader, Bryan Hanson, recalls, local authorities 'were moving like hell to keep up with everything'. He reflects today that 'the amazing thing is that the boom was for such a short period'. The question now is whether such conditions will recur, or whether alternative futures might emerge in and from an area which has seen enough of the social and environmental costs of industrial growth and decline.

References

Benn, C. and Fairley, J. (1986) *Challenging the MSC on jobs, training and education*, London: Pluto.

Beynon, H., Hudson R. and Sadler, D. (1986a) 'Nationalised industry policies and the destruction of communities: some evidence from north-east England', *Capital and Class* 29, pp. 27–57.

Beynon, H., Hudson R. and Sadler, D. (1986b) 'The growth and internationalisation of Teesside's chemicals industry' Middlesbrough Locality Study, Working Paper No. 3.

Briggs, A. (1963) *Victorian Cities*, Harmondsworth: Penguin.

Boswell, J. S. (1983) *Business Policies in the Making: Three Steel Companies Compared*, London: Allen and Unwin.

Carney, J., Lewis, J. and Hudson, R. (1977) 'Coal combines and interregional uneven development in the U.K., in D. Massey and P. Batey (eds), *Alternative Framework for Analysis*, London: Pion.

Cleveland County Council (1983) *The Economic and Social Significance of the British Steel Corporation to Cleveland*, Middlesbrough.

Cleveland County Council, (1987) *Unemployment Strategy*, Middlesbrough.

Fevre, R. (1986) 'Contract work in the recession', in H. Purcell, et al. (eds), *The Changing Experience of Employment: Restructuring and Recession*, London: Macmillan.

Financial Times (1981) 'A capital intensive cul de sac', 3 June.

Financial Times (1985) 'A town too proud to let the anger show', 19 March.

Foord, J., Robinson F. and Sadler, D. (1985) *The Quiet Revolution: Social and Economic Change on Teesside, 1965–1985*, Newcastle upon Tyne: BBC (NE).

Foord, J., Robinson F. and Sadler, D. (1987) 'Living with economic decline: Teesside in crisis', *Northern Economic Review*, 14, pp. 33–48.

Gladstone, F. (1976). *The Politics of Planning*, London: Temple Smith.

Globe and Mail, Toronto, 7 February 1987, 'England's great divide: the south prospers while the industries of the north crumble'.

Guardian (1983a) 21 March.

Guardian (1983b) 23 March.

Guardian, 1984 'Community without hope where the loan sharks prosper', 24 December.

Heath, E. (1985) 'Extracts from a speech given to the Sunderland Conservative Association, 14 January', mimeo.

HMSO (1963) *The North-east: A Programme for Regional Development and Growth*, London: HMSO, Cmnd. 2206.

HMSO (1985) *National Plan*, London: HMSO.

HMSO (1969) *Teesside Survey and Plan*, London: HMSO.

HMSO (1973) *British Steel Corporation: Ten Year Development Strategy*, London: HMSO, Cmnd. 5236.

HMSO (1978) *British Steel Corporation: The Road to Viability*, London: HMSO, Cmnd 7149

Hudson, R. and Sadler, D. (1984) *British Steel Builds the New Teesside? The Implications of BSC Policy for Cleveland*, Middlesbrough: Cleveland County Council.

Hudson, R. and Sadler, D. (1985) 'The development of Middlesbrough's iron and steel industry, 1841–1985, Middlesbrough Locality Study, Working Paper No. 2.

Labour Research Department (1939) *Coal Combines in Durham*, London: Farleigh.

Lewis, J. (1987) 'Employment trends in the financial services sector in Middlesbrough', Middlesbrough Locality Study, Working Paper No. 7.

Lock, M. (1946) *Middlesbrough Survey and Plan*.

Middlesbrough Evening Gazette, 30 May 1984.

NEDO (1986) *Changing Working Patterns: How Companies Achieve Flexibility to Meet New Needs*, London.

Nicholas, K. (1986) *The Social Effects of Unemployment in Teesside, 1919–39*, Manchester: Manchester University Press.

Pepler, G. and Macfarlane, P. W. (1949) *North-east Development Area Outline Plan*, Ministry of Town and Country Planning, interim confidential edition.

Pettigrew, A. (1985) *The Awakening Giant: Continuity and Change in ICI*, Oxford: Blackwell.

Priestley, J. B. (1934) *English Journey*, London: Heinemann.

Reader, W. J. (1970, 1975) *Imperial Chemical Industries: A History* (2 vols), London: Oxford University Press.

Roeber, J. (1975) *Social Change at Work: The I.C.I. Weekly Staff Agreement*, London: Duckworth.

Sadler, D. (1986) 'The impacts of offshore fabrication in Teesside', Middlesbrough Locality Study, Working Paper No. 5.

Smith, E. (1986) 'Living with unemployment in Cleveland', *Northern Economic Review* 13, pp. 2–17.

Sunday Times (1976) 'If only the speculators could see this'.

Townsend, A. (1986) 'Rationalisation and change in Teesside's health service', Middlesbrough Locality Study, Working Paper No. 4.

Townsend, P., Phillimore P. and Beattie, A. (1986) *Inequalities in Health in the Northern Region*, Newcastle on Tyne: Northern RHA.

9

The Local Question – Revivial
or Survival?

PHILIP COOKE

This book has presented an in-depth analysis of the ways in which
localities function internally and how they articulate to the wider
national and international processes of change in the political
economy. It has sought to illustrate the argument that the relationship
between the different scales is not simply a one-way street with
localities the mere recipients of fortune or fate from above. Rather,
localities are actively involved in their own transformation, though
not necessarily as masters of their own destiny. Localities are not
simply places or even communities: they are the sum of social energy
and agency resulting from the clustering of diverse individuals,
groups and social interests in space. They are not passive or residual
but, in varying ways and degrees, centres of collective consciousness.
They are bases for intervention in the internal workings of not only
individual and collective daily lives but also events on a broader
canvas affecting local interests.

Such propositions are an important outcome of the research
reported here. None of us thought in this way about localities at the
outset; indeed, much initial effort was devoted to establishing
whether or not the notion had any conceptual status at all. Nor
were the localities studied selected with the idea of examining just
how pro-active particular parts of the urban and regional system
had been or were being. The opposite would be closer to the truth:
the seven localities were selected to a considerable extent for their
variety. But this variety was determined much more by their
experiences in the wave of economic restructuring that affected UK
cities and regions in the 1970s and 1980s, their diverse labour
market and social structural trajectories, than by their local policy
performances. This is not to say that the last dimension was considered unimportant, rather that it was not a determining factor in
the choice of localities to be studied.

Accordingly, the discoveries regarding policy development, policy
initiatives and policy failure that have been described can be taken as

Table 9.1 *Classification of policy pro-activity in the seven study localities*

	Pro-active	Non-pro-active
Political	Middlesbrough −	Thanet + Cheltenham +
Managerial	Swindon + Outer Liverpool +	South West Birmingham − Lancaster +

± = Pro-activity trend

not untypical of the range of policy postures being adopted by UK localities having middle of the road political cultures. As such, they are interesting for the light they shed on the different emphases and social origins of the policies being pursued currently and in the relatively recent past. This changing diversity is schematized in Table 9.1 which contrasts traditional with contemporary policy dispositions and sources of initiative in the seven study localities. Three general deductions can thus be made on the basis of the evidence drawn from the seven localities.

The first is that there is a clear division between localities having a history of active, local policy intervention and those lacking such a history. It is extremely important to recognize that such intervention is not confined to economic development issues; local policy effectiveness is often focused upon social care. Moreover, local initiative frequently occurs outside the formal sphere of local government. For the present, unfortunately, employment creation has the higher local profile, a characteristic which has, perforce, to be reflected here. Even now, however, diverse emphases influenced by local traditions continue to condition local priorities. The relevant chapters in the book explain these distinctive policy histories. In very simplified terms the difference can be boiled down to paternalism and the struggle to escape it. Paternalism is too complex a concept to explore in detail here: it can be active and stifling of popular initiative as it was at the Lancaster factories or at Cadbury's in Birmingham, or passive and enervating as in Cheltenham and Thanet where it had its basis as much in local politics as in the workplace. By contrast, those localities with a history of intervention had, to some extent, sloughed off paternalistic tradition, and not necessarily solely by means of local effort. For example, both Middlesbrough and Swindon were classic company towns with all that is implied by that label for control of local social and political life. But the long-term decline in power of local industrialists and, more importantly perhaps, nationalization by the central state of key industries led to the resulting vacuum being occupied by the local labour movement. Accordingly, more innovative kinds of local interest representation, particularly towards the central state, became feasible.

The second general feature of the local policy process is that its vitality waxes and wanes over time and space. Thus the urban–industrial localities – the modern formation of which was closely tied to the Fordist era with its emphasis on mass production, mass housing, welfare provision and, to varying degrees, state regulation of the development process – are now less obviously interacting in their own restructuring processes than they were. The outer suburbs of Liverpool centred upon Kirkby, Halewood and Speke were largely formed by the interaction of central state regional policy redirecting the location of, especially, motor vehicle production and local state housing policy as part of a wider restructuring plan for the Greater Liverpool area. Birmingham's motor industry expansion was less dependent upon central state policy of the kind responsible for Liverpool's transformation, but the provision of mass housing close to the booming peripheral industrial areas was a central feature of the city's urban redevelopment in the postwar years. And Middlesbrough, confronted with industrial decline in long-established industry, mobilized its local coalition successfully to press government and large-scale industry for new infrastructural and industrial investment. But in the context of a shift in the mode of developmental organization in which large-scale industry, the mass provision of public sector housing and state regulation of development are no longer dominant motifs, that kind of local initiative and effectiveness has become redundant.

That is, of course, not to say that local pro-activity is no longer relevant as a general rule. Despite the undoubted increase in discretionary control of local affairs by the central state in the 1980s, there is substantial evidence that localities continue to mobilize in a diversity of ways to cope with change. In particular, local policy is now more alert to the manner in which local effectiveness depends on the extent to which footholds can be formed to link the local to the global. At one extreme can be found the phenomenon of 'place marketing' whereby whatever is idiosyncratic or unique about a particular locality is packaged and sold to the outside world as a commodity without which, whether as tourists or international investors, they will be unnecessarily depriving themselves (for critiques, see Wright, 1985; Horne, 1986; Hewison, 1987). At another extreme could be found, until the 1986 abolition of metropolitan counties, the example of the Greater London Council, developing for the first time a fully researched industrial strategy for coping with the massive job loss occasioned by economic restructuring (Greater London Council, 1985). In this, links were formed, through the adoption of a municipal trade policy, with developing countries such as Nicaragua and Vietnam, and through international labour organizations, with trade unions in multinational companies̆having a base in the London economy as well as those of cities in other countries.

Initiatives such as these can require a prominent, not to say leading, involvement by officers, especially where political leadership may be so ambitious as to have outreached the inherited norms of policy practice or, conversely, where a local political regime has yet to adjust fully to a rapidly changing policy environment. Our research has shown, thirdly, that where political vision is lacking, clear-sighted managerial initiative both inside and, sometimes, outside the local policy arena can be crucial. In Lancaster, for example, neat footwork by the town clerk's department ensured that land shortage would not be an obstacle to the possible location of the university in what at the time was a declining local industrial economy. Accordingly, when the university arrived, it brought in new personnel and new ideas that, in the persons of a modernizing intelligentsia, helped further strengthen local initiative, especially in the conversion of old buildings for new industry and the general beautification of the built environment. The urban managerial influence upon Swindon in the past has been written about elsewhere (Harloe, 1975) and it is a tradition that continues in one of the country's most assiduous exercises in the place marketing of a high technology milieu. But in other localities the initiative has come from within the political system, notably in Thanet, where stagnation and the decline of tourism have resulted in local interests pressuring the council into sometimes reckless policy formulation.

However, though there is plenty of evidence of the emergence of local pro-activity, ranging from local boosterism to local chauvinism, it is also clear that the pay-off in terms of the ultimate criterion by which such activity demands to be measured – job creation – is rather small (Benington, 1986). Estimates, subject to much debate as to reliability, suggest that even some of the largest UK metropolitan Enterprise Boards could claim only relatively modest job creation totals by comparison with, for example, the Scottish and Welsh Development Agencies funded by the central state. It is difficult to get an accurate assessment of the numbers of jobs created by either Enterprise Boards or Development Agencies. The latter usually report on 'job opportunities' created or 'jobs forecast' and a strong case can be made for dividing such estimates in half, at least, to arrive at a more credible estimate (Cooke, 1987). However, even if that division is performed it seems likely that the Scottish and Welsh Development Agencies have been generating between 2000 and 3000 jobs each *per year* since 1976 (Moore and Booth, 1986a; Welsh Development Agency, 1981, 1986).

A key factor, though not the only important one, in explaining the difference in performance is the massive contrast in the levels of expenditure available to the Enterprise Boards by comparison with the Development Agencies. The Welsh Development Agency began with a budget of £20 million per year in 1976 and was, ten years later, spending £60 million annually on its various activities. The Scottish Development Agency started with £60 million per year and was averaging over £100 million expenditure per year by the mid-1980s (Moore and Booth, 1986b). To this essential factor can be added the organizational capability that these agencies have developed over a decade, so that they are now able to target investment increasingly towards what are perceived as growth sectors, rather as Japan's Ministry for International Trade and Industry (MITI) has successfully done on a much vaster scale. Moreover, they are able to represent their countries' interests overseas, thereby attracting inward investment. By contrast, Enterprise Boards tend to have been tied to investing in troubled companies and pursuing socially useful and desirable goals – something which has not featured prominently in development agency strategies in recent years.

There is a case for building upon the strong signs of local pro-activity revealed, even for quite small localities, in the research reported here. There is also evidence that if adequate investment capital is available, as it is for the Scottish and Welsh Development Agencies, then not insignificant contributions can be made to the job-generation process. The 50,000 new jobs ascribable conservatively to agency activity since 1976 cannot be dismissed as a drop in the ocean. The question that has to be addressed is how best might local initiative be harnessed to an organizational structure capable of focusing clearly on a precise target, such as the achievement of a set job increment, within given cost constraints and without neglecting local democratic rights and sensibilities? Despite appearances this is not necessarily the scenario for an episode of *Mission Impossible*.

There are four possible models of policy delivery of this kind that are worth considering. They are respectively:

1   municipal enterprise;
2   public development corporations;
3   regional–local partnerships;
4   autonomous development authorities.

Each is potentially a broadly equivalent means of delivering large sums of revenue to specific projects in a context where market failure has resulted in blocked local development. Crucially, however, each requires access to the funds necessary to enable projects to be accomplished. The key differences between them concern, first, their means of raising necessary funding and, second, their means of administration.

Municipal enterprise already exists but in a relatively weakened situation in a political climate hostile to public initiative and, it could be said, the very existence of a public sector more generally. It usually, though not exclusively, involves the establishment of a quasi-autonomous enterprise board accountable to democratically elected municipal representatives. As has been noted, municipal enterprise of this kind has been hamstrung primarily by limitations on revenue-raising capacity. Most have been dependent upon the product of a two pence rate and whatever extra capital that can be raised from, for example, allocating a small percentage of the municipal pension fund to the general investment pool. Latterly, however, enterprise boards that have operated a low political profile have been able to raise much larger sums on the open finance market. A good example of this is Lancashire Enterprises Ltd (LEL), particularly active, as we have seen, in Lancaster. LEL raised £10 million in private loans in 1986 (Hetherington, 1987). It has emphasized investing in new, small firms rather than the larger, failing ones supported by the Greater London or West Midlands Enterprise Boards and, as a consequence, it has been able to present an attractive face to the notoriously antipathetic private banking sector. Nevertheless, at £4 million, LEL's annual capitalization remains small and its impact, though locally sensitive, targeted and democratically accountable, is likely to be marginal even in the relatively small localities like Lancaster and Wigan where it has been most active.

Public development corporations also exist and have become, in the form of Urban Development Corporations (UDCs), the main means of delivering significant urban investment to declining inner city areas under the Thatcher government. Unlike municipal enterprise, UDCs are in no meaningful way democratically accountable, other than through the presiding cabinet minister in Parliament. However, also unlike municipal enterprise, UDCs are supplied with appropriate levels of funding, sufficient to make the all-important contribution to turning very large, derelict inner-city areas into attractive investment opportunities for private capital. The public capital with which UDCs subsidize private investment is sourced from the Treasury allocation to the Department of the Environment or the Welsh and Scottish Offices, and so is ultimately a product of central state tax revenue. The level of expenditure is thus solely at the discretion of the minister responsible. One of the main criticisms of

UDCs is that they represent a totally centralist solution to urban problems and one, moreover, which under a right-wing government rests on transforming deprived working class neighbourhoods into middle class playgrounds composed of yachting marinas, expensive shopping plazas and elite culture facilities, notably opera houses to service up-market residential and tourist accommodation. The levels of local sensitivity and democratic accountability are thus low in the UDCs but they are relatively successful instruments for manufacturing attractive neo-heritage environments. This is principally because of the relatively generous funding at their disposal.

Regional–local partnerships have had no significant presence in the UK principally because of the highly centralized tradition of administering 'the regions', especially in England. Even in Northern Ireland, Scotland and Wales, where greater administrative and economic development autonomy exists, development agencies tend to intervene autonomously – provide, for example, advance factories suitable for private sector occupancy, and allocate tenants accordingly. As a consequence the prospects for local integration between co-occupiers of a given industrial estate are absolutely minimized. The concept of 'partnership' has specific connotations in British urban policy, first reflecting the links between central and local government in 1970s Inner Cities policy, then more recently denoting the enhancement of public sector links with private sector property development interests under the 1980s Urban Development Corporation and other Urban Programme legislation. However, relatively little of this type of activity has matched the success of the less trumpeted but more effective forms of partnership associated with local economic and employment growth in parts of Italy. The 'third Italy' as the area between the industrial north-west and the Mezzogiorno has come to be called (Brusco, 1982; Cooke, 1984; Murray, 1987) has moved from being a relatively backward part of the country to being its most productive, efficient and, for some, wealthy. As Sabel (1982) shows, no small part in the success of one part of this area, the region of Emilia-Romagna centred on Bologna, can be ascribed to the following:

(a) The existence of a shared (Leftist) political culture between regional and local politicians on the one hand, and the constellation of small business-owners on the other.
(b) The extensive aid given jointly by regional and local agencies to small employers, including widespread provision of infrastructure, industrial parks, etc.
(c) The advice provided for employers by the regional government through its specialist agencies on market strategy, business information, industry analyses and the use of technology.

(d) Special grant-aid to cooperative enterprise to enable producers also to become distributors, thus cutting out the middleman.

(e) Regional and local government placing of local orders.

In the early 1980s proposals for collective provision of accountancy services, marketing opportunities, workforce training and technology assessment were also under active consideration. Such activity is not confined to radical or 'red' regions such as Emilia: it exists, albeit in a more entrepreneurial form, in 'white' or Catholic regions such as Veneto where the highly successful Benetton clothing company is based. Some of these ideas were adopted by the Greater London Council and other local authorities (Murray, 1985a; 1985b; 1985c; Nolan and O'Donnell, 1987; Davenport and Totterdill, 1986) but progress has been handicapped by expenditure constraints.

In the UK in the late 1980s expenditure constraints limit the freedom of manoeuvre available to municipalities to an extent seldom experienced previously. The significant exceptions are the Urban Development Corporations earmarked to spend £126 million in 1987/8. Other centralist initiative such as the Urban Programme (£294 million in 1987/8), Urban Development Grants (£120 million spent 1983–1987) and sundry urban expenditures from Derelict Land Grant to Enterprise Zones mean that significant sums are being spent but that what it is being spent on is determined in London rather than by the elected representatives of the recipient localities. This is unhealthy in that it threatens to impose a metropolitan perception of the appropriate use of inner city land upon the residents of cities who may not share such a perspective. A key question is whether a policy instrument exists which would allow localities to raise large sums of finance for specific, locally determined projects without infringing Thatcher government ratecapping or exchequer-transfer restrictions. One which has a long history of success in the USA is the autonomous development authority.

The success of autonomous development authorities arose from their ability to raise funds through issuing bonds secured by the revenues received from projects implemented (Lines, Parker and Perry, 1984). The most celebrated examples of these authorities have included the Port of New York Authority and the Triborough Bridge Authority which Austin Tobin and Robert Moses used to modernize the communication system of the city and its environs (Berman, 1983). Local councils in the UK can, of course, issue bonds but the process does not have a history of being employed to finance specific development projects by means other than borrowing from banks, raising property taxes or receiving central state stransfers. It is likely that local government in the UK is debarred from establishing bond-issuing autonomous development authorities and, in any case, with local government relatively well funded from its own sources and central transfers in the past, there was no obvious need for it. However, in an era of tight central control on the nature and content of local authority budgets, combined with a neoliberal espousal of the virtues of the marketplace as a determinant of social and economic change, the time may have arrived when new means of financing developments which reflect local rather than central preferences deserve consideration. The autonomous development authority might be one among many worthy candidates.

Hence, in terms of the question posed earlier regarding the prospect for establishing targeted, locally sensitive and economical means of achieving local developmental success, it is clear that a variety of modes can be identified which are partly satisfactory. The resolution of the question is closely bound up with national and local political ideology too, but that local energy, initiative – pro-activity – should be penalized is presumably undesirable from any political perspective. Under a less centralist political regime at national level it would seem wise to encourage the impressive organizational advances represented in the Enterprise Board concept by significantly enlarging the budgetary base of such organizations. They come close to the successful localist developments in the third Italy in some ways, though they would benefit from the efficiencies that could accrue from the provision of certain services at a supra-local, probably regional, or at least development authority level of government. This might work best outside England, which lacks strong regionalist traditions and has, instead, historically strong cities and counties – some of which have proved to be innovators in the job generation process, although severely hamstrung by financial constraints.

If national government in the UK remains under the control of neoconservative interests, they too will be confronted sooner or later by the limits of a method of administration which involves centralizing power while devolving responsibility. The difficulties of tightening the National Health Service budget in real terms while blaming its employees for the resulting problems were eventually to provoke resistance and protest occasioning a review of policy. Giving back local autonomy while tightly gripping the fiscal purse-strings is an impossibility. And running the urban affairs of an increasing number of localities from London is no way of constructively responding to local interests and concerns over the long term. While Urban Development Corporations, adequately funded, can in the appropriate locational context transform disadvantaged areas of cities into middle class havens with surprising speed, they are far from universally popular. This is not least because of the way in which centralist power used negatively to control municipal enterprise can leave some localities looking feeble and helpless. It thus helps create an apathetic or dependent, clientilist culture, something precisely at odds with the professed Thatcherite project. So within even that particularist perspective there is a case for encouraging innovation in local revenue raising and policy development.

At present, it can be argued, government policy towards the electorate, dispersed as it is in the multifold localities of the UK, is too restrictive in a downward direction and too liberal upwards and outwards to a global economy which is subject to increasingly alarming fluctuations. These chapters have shown just how important shifts in the global economy can be for the fate of localities. The Thatcher governments have further exposed to the gales of international competition an economy that was already one of the world's most open. But there is a responsibility on government to seek international agreements on trade and currency exchange rates of the kind that existed during the postwar boom years. Bretton Woods and the General Agreement on Tariffs and Trade (GATT) became too rigid for the flexible times of the 1970s. But the late 1980s have begun to reveal the limitations of international trading and exchange rate flexibility. A society which is exposed to the presently unregulated forces of an increasingly globalized economy while being deprived of discretion over local affairs, some of which involve picking up the pieces left by global economic whirlwinds, is Promethean in its predicament. The Prometheus myth, it is worth recalling, ended with the opening of Pandora's box.

References

Benington, J. (1986) 'Local economic strategies', *Local Economy*, 1, pp. 7–24.

Berman, M. (1983) *All That is Solid Melts Into Air*, London: Verso.

Brusco, S. (1982) ' "The Emilian model": productive decentralization and social integration', *Cambridge Journal of Economics*, 6, pp. 167–184.

Cooke, P. (1984) 'Region, class and gender: a European comparison', *Progress in Planning*, 22, pp. 87–146.

Cooke, P. (1987) 'Wales', in P: Damesick and P. Wood (eds), *Regional Problems, Problem Regions and Public Policy in the United Kingdom*, Oxford: Clarendon Press.

Davenport, E. and Totterdill, P. (1986) 'Fashion centres: an approach to sector intervention', *Local Economy*, 1, pp. 57–63.

Greater London Council, (1985) *The London Industrial Strategy*, London: GLC.

Harloe, M. (1975) *Swindon: A Town in Transition* London: Heinemann.

Hetherington, P. (1987) 'Only winners will do . . . ' *The Guardian*, 26 November, p. 33.

305

Hewison, J. (1987) *The Heritage Industry*, London: Methuen.

Horne, D. (1986) *The Public Culture*, London: Pluto.

Lines, J., Parker, E. and Perry, D. (1984) 'Building the twentieth century public works machine: Robert Moses and the public authority', *Studies in Planning and Design*, Buffalo: State University of New York.

Moore, C. and Booth, S. (1986a) 'The Scottish Development Agency: market consensus, public planning and local enterprise', *Local Economy*, 3, pp. 7–19.

Moore, C. and Booth, S. (1986b) 'Unlocking enterprise: the search for synergy', in W. Lever and C. Moore (eds), *The City in Transition*, Oxford: Clarendon Press.

Murray, F. (1987) 'Flexible specialisation in the "Third Italy" ', *Capital and Class*, 33, pp. 84–96.

Murray, R. (1985a) 'New directions in municipal socialism', in B. Pimlott (ed.), *Fabian Essays in Socialist Thought*, London: Heinemann.

Murray, R. (1985b) 'London and the GLC: restructuring the capital of capital', *IDS Bulletin*, 16, pp. 47–55.

Murray, R. (1985c) 'What are the lessons from London?', in K. Coates (ed.), *Joint Action for Jobs: A New Internationalism*, Nottingham: Spokesman.

Nolan, P. and O'Donnell, K. (1987) 'Taming the market economy? A critical assessment of the GLC's experiment in restructuring for labour', *Cambridge Journal of Economics*, 11, pp. 257–264.

Sabel, C. (1982) *Work and Politics*, Cambridge: Cambridge University Press.

Welsh Development Agency (1981) *The First Five Years*, Cardiff: WDA.

Welsh Development Agency (1986) *Annual Report and Accounts*, Cardiff: WDA.

Wright, P. (1986) *On Living in an Old Country*, London: Verso.

Index

307